INTRODUCTION TO
NUMERICAL METHODS FOR PARALLEL COMPUTERS

ELLIS HORWOOD SERIES IN MATHEMATICS AND ITS APPLICATIONS

Series Editor: Professor G. M. BELL, Chelsea College, University of London

Statistics and Operational Research

Editor: B. W. CONOLLY, Chelsea College, University of London

Beaumont, G. P.	**Introductory Applied Probability**
Beaumont, G. P.	**Basic Probability and Random Variables***
Conolly, B. W.	**Techniques in Operational Research: Vol. 1, Queueing Systems**
Conolly, B. W.	**Techniques in Operational Research: Vol. 2, Models, Search, Randomization**
French, S.	**Sequencing and Scheduling: Mathematics of the Job Shop**
French, S.	**Decision Theory**
Hartley R.	**Linear Methods of Mathematical Programming***
Jones, A. J.	**Game Theory**
Kemp, K. W.	**Dice, Data and Decisions: Introductory Statistics**
Oliveira-Pinto, F.	**Simulation Concepts in Mathematical Modelling**
Oliveira-Pinto, F. & Conolly, B. W.	**Applicable Mathematics of Non-physical Phenomena**
Schendel, U.	**Introduction to Numerical Methods for Parallel Computers**
Stoodley, K. D. C.	**Applied and Computational Statistics A First Course**
Stoodley, K. D. C., Lewis, T. & Stainton, C. L. S.	**Applied Statistical Techniques**
Thomas, L. C.	**Games, Theory and Applications**
Whitehead, J. R.	**The Design and Analysis of Sequential Clinical Trials**

**In preparation*

INTRODUCTION TO NUMERICAL METHODS FOR PARALLEL COMPUTERS

U. SCHENDEL
Professor of Numerical Mathematics and Computer Science
Freie Universität (Free University) of Berlin

Translator:
B. W. CONOLLY
Professor of Operational Research
Chelsea College, University of London

ELLIS HORWOOD LIMITED
Publishers · Chichester
Halsted Press: a division of
JOHN WILEY & SONS
New York · Chichester · Brisbane · Toronto

First published in 1984 by
ELLIS HORWOOD LIMITED
Market Cross House, Cooper Street, Chichester, West Sussex, PO19 1EB, England

The publisher's colophon is reproduced from James Gillison's drawing of the ancient Market Cross, Chichester.

Distributors:
Australia, New Zealand, South-east Asia:
Jacaranda-Wiley Ltd., Jacaranda Press,
JOHN WILEY & SONS INC.,
G.P.O. Box 859, Brisbane, Queensland 40001, Australia
Canada:
JOHN WILEY & SONS CANADA LIMITED
22 Worcester Road, Rexdale, Ontario, Canada.
Europe, Africa:
JOHN WILEY & SONS LIMITED
Baffins Lane, Chichester, West Sussex, England.
North and South America and the rest of the world:
Halsted Press: a division of
JOHN WILEY & SONS
605 Third Avenue, New York, N.Y. 10016, U.S.A.

British Library Cataloguing in Publication Data
Schendel, U.
Introduction to numerical methods for parallel computers. –
(Ellis Horwood series in mathematics and its applications)
1. Numerical analysis – Data processing
2. Parallel processing (Electronic computers)
I. Title II. Einfuhrung in die parallele Numerik. *English*
519.4 QA297

Library of Congress Card No. 84-10941

ISBN 0-85312-597-X (Ellis Horwood Limited)
ISBN 0-470-20091-X (Halsted Press)

Typeset by Ellis Horwood Limited.
Printed in Great Britain by R.J. Acford, Chichester

Table of Contents

Preface to the English Edition

In offering an English translation of the German original both author and translator hope that a wider audience for this fascinating subject will be captured. It seems still to be true that, whereas hardware descriptions of the advanced modern machines with which this book is concerned have a growing literature, the problems posed by the need to create a suitable body of accompanying mathematical software are not being tackled and documented quite so systematically. The subject invites the attention of all who are interested in the improved computational possibilities afforded by modern technology and we hope that the book will provide an introduction as well as documentation of existing methodology.

The opportunity has been taken to add to and expand some of the material as well as to correct misprints and errors. The additions include an enlarged treatment of eigenvalue procedures in Chapter 4, and two Appendices containing material which was not available when the book was written. The first Appendix supplements Chapters 1 and 2 by providing technical information about a number of particular machines and their performance characteristics in certain benchmark tests. The second Appendix gives a brief account of some parallelised algorithms for nonlinear optimisation.

Translator and author have enjoyed a pleasant and fruitful collaboration which has established a friendship as well as providing a rewarding professional experience. The author is very grateful to Professor B. W. Conolly for the translation of the book and for his improvements and to his colleagues Dipl. Math. J. Brandenburger and Dipl. Math. M. Schyska for reading the manuscript.

Berlin U. Schendel
London B. W. Conolly
1984

Foreword

This book is the result of a series of lectures at the Free University of Berlin, intended to provide an introduction to the principles of parallel numerical analysis. This branch of numerical mathematics has received its impetus and significance from the development of new computer architectures, in particular from the concept of the parallel computer, which has made it possible to tackle larger and more complex numerical problems than hitherto. The procedures available for the solution of such problems must often be 'parallelised' in such a way as to guarantee their successful operation on the corresponding parallel computer.

In this book certain recognised principles are set forth for the development of parallel numerical algorithms. Because no standard computer model for use in the development of parallel numerical methods in a unified theory has yet been developed, so far the various concepts proposed tend to depend heavily on the particular class of computer under discussion. The choice of material is confined by and large to those numerical procedures which form a part of the body of classical and well-tried methods in numerical analysis. On the other hand possible ways are discussed for the development of parallel algorithms for particular problems.

Primary targets of this book are the development of a fundamental understanding of the principles of parallel numerical methods and preliminary guidance in the solution of problems. The mathematical apparatus is presented in such a way that scientists and engineers alike can be initiated into the subject.

I am grateful to Prof. Dr. Feilmeier (Braunschweig) and Prof. Dr. Sameh (Urbana, illinois) for many stimulating discussions, and to my former colleague Dipl. Math. K. Kupfernagel for his contributions and critical remarks. Finally I wish to express my special indebtedness to the Oldenbourg Verlag for their interest and excellent collaboration.

Berlin 1981 U. Schendel

1

Introduction

The model currently used to describe numerical methods in the context of digital computers is the well-known universal model of von Neumann. This is illustrated in Fig. 1.1.

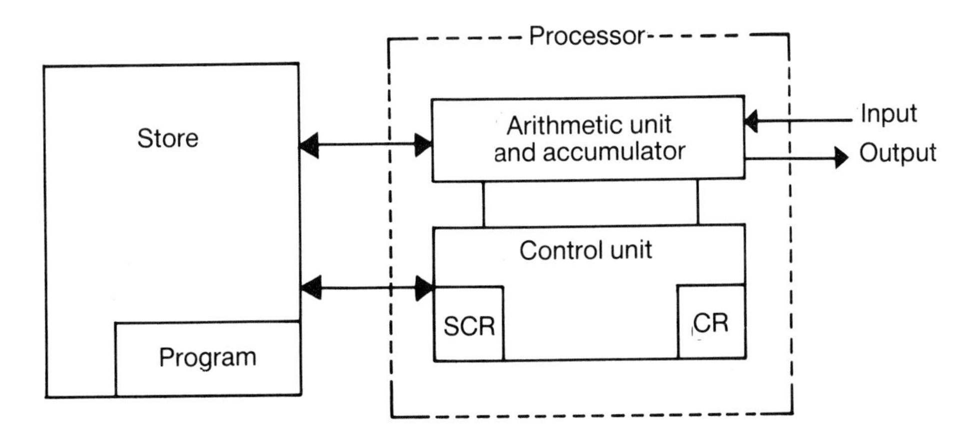

Fig. 1.1 – Universal computer

Legend

SCR = Sequence Control Register (contains the address of the next instruction).

CR = Control Register (contains a copy of the current instruction).

The universal computer possesses the following features:

(1) digital representation of variables;
(2) processing is carried out serially according to the basic operations of arithmetic and logic;
(3) the program is a coded version of the algorithms to be implemented and the data are held in the main store.

All further developments have taken this basic model as their starting point. In particular, the algorithms of numerical analysis are based on this type of computer model and entail a large number of elementary operations. Problems of rounding errors and their propagation, and also questions of numerical stability, turn out to be of great importance [1].

The performance of serial machines has improved enormously during the last twenty years as a result of new technology and improved design. Some features of parallelism have been introduced. These include:

(1) the organisation of input/output channels;
(2) overlapping in the execution of instructions;
(3) interleaved storage techniques.

These developments have nevertheless lagged behind the requirements of numerical methods, even at the programming level.

In the 1960s the possibility of truly parallel digital computers began to be discussed. Parallelism was to be interpreted in the widest sense, and the objective was to offer the user the full capability of such machines. Computers of this kind have indeed been available for some time, examples being the CDC STAR 100, the ILLIAC IV, STARAN, the CRAY-1, Texas Instruments TI ASC, IBM 2938, IBM 360/91, CDC 7600. However, there remains a need for a 'parallel numerical analysis', that is, a body of numerical mathematics which takes specific advantage of the possibilities offered by parallel computers. Of greater theoretical interest is the question of whether there exists a *maximal parallelism* for a given range of problems. Questions of this nature belong to the domain of complexity theory.

There follows a list of some of the pros and cons of parallel computers.

PRO

(1) Increase of computing speed
    (a) conventional structures are at their physical limits
    (b) a single computer with n processors is cheaper than n computers each having a single processor.
(2) Possibility of solving problems still too complex for serial machines. World-wide weather forecasting provides an example.
(3) The solution of problems which are by nature parallel. Operations with vectors and discrete simulation provide examples.
(4) Real time problems. Examples are provided by the processing of typographical data (picture and graphic processing), aerial survey.

CON

(1) Poor utilisation of the machine (a management problem).
(2) Complicated organisation of the data (parallel access).

Great difficulty arises from the fact that at present there is no standard model for parallel systems. This matter will be considered further in Chapter 2 when we examine what are called SIMD and MIMD models.

It is of the utmost importance in parallel numerical analysis to be able to assess the speed gain expected from the operation of p processors in parallel. For this purpose an operational measure S, called the *speed-up* ratio, is introduced. It is defined by

$$S = \frac{\text{Computing time on a serial machine}}{\text{Computing time using a parallel machine}}. \tag{1.1}$$

To obtain a valid result it is essential, for a given problem, to compare the *best* serial algorithm available with the *best* parallel algorithm, even if these are different. According to Stone [2] S has essentially four possible forms:

S	*Examples*
(a) $S = kp$	Matrix calculations, discretisation.
(b) $S = kp/\log_2 p$	Sorting, tridiagonal linear systems, linear recurrence formulae, evaluation of polynomials.
(c) $S = k \log_2 p$	Searching.
(d) $S = k$	Certain nonlinear recurrences and compiler operations.

Here k is a machine-dependent quantity such that $0 < k < 1$ (and is near to unity), and p is the number of processors. The classification of particular types of problem according to one of (a) to (d) is, however, provisional until they can be identified within the framework of complexity theory. The performance limits attainable by a reasonable parallel numerical procedure lie between (b) and (c). The computing time required for an algorithm for a parallel computer is both unknown and machine-dependent. To be able to assess the merits of a parallel algorithm one needs to count the number of time unit steps needed. For this purpose it is convenient to introduce the idealised notion that during such a time unit step exactly one arithmetical operation can be carried out in the parallel mode (see Chapter 4). Let T_p be the number of time unit steps required by a parallel algorithm designed to utilise p ($\geqslant 1$) processors. Then T_1 is the time needed by the corresponding serial algorithm. In order to measure the attributes of parallel operation the following parameters are introduced.

Definitions
The *speed-up* factor of a parallel algorithm by comparison with its serial counterpart is defined by

$$S_p = T_1/T_p \quad (\geqslant 1),$$

(see also 1.1), and the *efficiency* by

$$E_p = S_p/p \quad (\leqslant 1). \tag{1.2}$$

E_p measures the *utilisation* of the parallel machine. The longer processors are idle, or carry out extra calculations introduced through the parallelisation of the problem, the smaller becomes E_p.

To compare two parallel algorithms for the same problem the following measure of effectiveness F_p is introduced:

$$F_p = S_p/C_p \ , \tag{1.3}$$

where

$$C_p = pT_p \tag{1.4}$$

measures the 'cost' of the algorithm. Note that

$$F_p = S_p/(pT_p) = E_p/T_p = E_pS_p/T_1 \leqslant 1 \ . \tag{1.5}$$

F_p is thus a measure both of speed-up and efficiency. A parallel algorithm can accordingly be regarded as effective if it maximises F_p.

The following simple example is given for illustration. Suppose that it is required to form the sum

$$A = \sum_1^{16} a_i \ .$$

To carry this out with a single processor sequentially requires 15 additions. If we take the number of additions as a measure of the time needed, then normalising by assuming the time for a single addition to be the unit, we have $T_1 = 15$. If two processors were available we would form the two sums

$$b_1 = a_1 + a_2 + \ldots + a_8 \ , \qquad b_2 = a_9 + a_{10} + \ldots + a_{16}$$

simultaneously, requiring seven time units, and then form

$$c = b_1 + b_2$$

at the next stage, requiring a further time unit. Thus A would be obtained in $T_2 = 7 + 1 = 8$ time units. If now we had three processors A could be formed in the following three stages:

$$b_1 = a_1 + a_2 + a_3 + a_4 + a_5, \quad b_2 = a_6 + a_7 + a_8 + a_9 + a_{10},$$
$$b_3 = a_{11} + a_{12} + a_{13} + a_{14} + a_{15};$$
$$c_1 = b_1 + b_2, \quad c_2 = b_3 + a_{16};$$
$$d_1 = c_1 + c_2 = A \ .$$

This requires $T_3 = 4 + 1 + 1 = 6$ time units. Given four and eight processors the following procedures and their corresponding times are feasible.

$p = 4$

$$\left.\begin{aligned} b_1 &= a_1 + \ldots + a_4, \quad b_2 = a_5 + \ldots + a_8, \quad b_3 = a_9 + \ldots + a_{12}, \\ b_4 &= a_{13} + \ldots + a_{16} \end{aligned}\right\} \quad (3)$$

$$c_1 = b_1 + b_2, \quad c_2 = b_3 + b_4 \quad (1)$$

$$d_1 = c_1 + c_2 \quad (1)$$

$$T_4 = 5.$$

The numbers in ( ) mean the needed time units.

$p = 8$

$$b_1 = a_1 + a_2, \quad b_2 = a_3 + a_4, \ldots, b_8 = a_{15} + a_{16} \quad (1)$$

$$c_1 = b_1 + b_2, \quad c_2 = b_3 + b_4, \ldots, c_4 = b_7 + b_8 \quad (1)$$

$$d_1 = c_1 + c_2, \quad d_2 = c_3 + c_4 \quad (1)$$

$$A = d_1 + d_2 \quad (1)$$

$$T_8 = 4.$$

A table of the performance measures can now be constructed

p	T_p	C_p	S_p	E_p	$F_pT_1=S_pE_p$
1	15	15	1	1	1
2	8	16	1.88	0.94	1.76
3	4	18	2.5	0.83	2.08
4	5	20	3	0.75	2.25
8	4	32	3.75	0.47	1.76

The table shows that with increasing p, S_p increases steadily while E_p decreases. F_pT_1, however, has a maximum when p=4 which indicates that p=4 is the optimal choice of number of processors for this calculation. However, as we saw, the algorithms are different in detail for each p and the variations in S_p, E_p and F_p are partly the consequence of different parallel algorithms for the same problem. The parameters introduced above give one measure for the assessment of a parallel algorithm. Other aspects for consideration are stability and the analysis of errors. Of the greatest importance, however, is to recognise which problems already possess a parallel character, and which can be 'parallelised'.

2

Possible Computer models

Considerable difficulties surround the problem of formulating a standard machine model for parallel numerical methods, yet such a model is needed for theoretical purposes as well as to provide a background for the the development of algorithms.† The situation is quite different from that met in classical numerical analysis where von Neumann's universal computer model provides a basis. Figure 2.1 shows a general model of a parallel computer. This is, however, still too general for detailed statements to be made concerning the functioning of the computer. In particular it is too general for the development of algorithms. For further complementary discussion see Appendix 1.

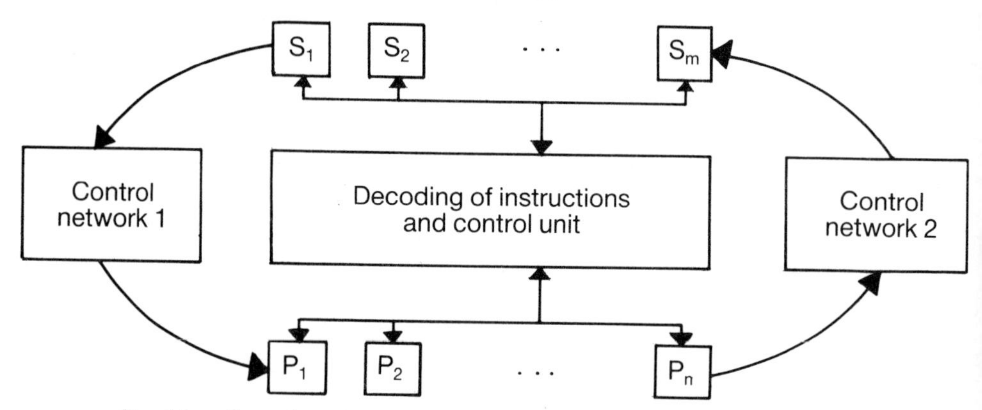

Fig. 2.1 – General configuration of a parallel computer with different levels of Parallelism.
S: stores; P: processors

Note: Parallelism is possible
(1) within the control unit;
(2) among the processors;
(3) among the stores;
(4) in the data central networks.

† Useful discussions of modern high performance machines are given by Ibbett [3], Zakharov [50] and Hockney [51]. This is, however, somewhat technical for a beginner.

We can now introduce an important classification of computers due to Flynn [4]. First we consider the number of streams of instructions and classify machines according to these into types:

$$\begin{aligned} &\text{SI} = \text{Single instruction stream} \\ &\text{MI} = \text{Multiple instruction stream} \end{aligned} \qquad (2.1)$$

Next a distinction is made according to the number of streams of operands:

$$\begin{aligned} &\text{SD} = \text{Single data stream} \\ &\text{MD} = \text{Multiple data stream} \end{aligned} \qquad (2.2)$$

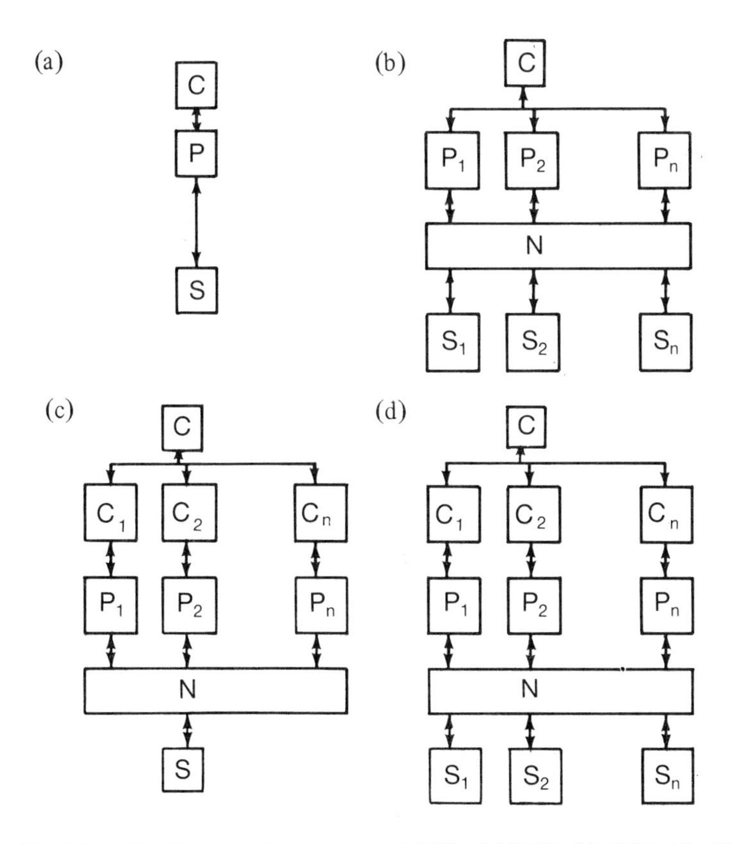

Fig. 2.2 – Classification of computers. (a) SISD. (b) SIMD. (c) MISD. (d) MIMD. (C: control unit; P: processor; N: data organisation network; S: store).

Thus, four combinations are possible:

SISD: This is the classical von Neumann model. A single stream of instructions operates on a single data stream.

SIMD: This is the class to which array processors and pipeline processors belong. All the processors interpret the same instructions and execute them on different data. Because of their simple form machines of this kind can have a large number of processors. For instance, the ICL/DAP has 4096.

MIMD: This is the multiple processor version of SIMD. All processors interpret different instructions and operate on different data. Because of the resulting complexity it is not surprising that the number of processors is usually small, at the time of writing up to 16 (DENELCOR HEP).

MISD: (Chains of processors). It can be argued (see Appendix 1) that this type is equivalent to SISD and is, therefore, of no importance in this book.

These machine types are illustrated by Fig. 2.2.

2.1 SIMD processors

According to their architecture and method of functioning machines of this type are classified into the following types:

(i) *Array processors:* for example the Burroughs ILLIAC IV.
(ii) *Pipeline processors:* also called vector processors. Examples are the CDC STAR-100, Texas Instruments ASC (Advanced Scientific Computer), CRAY-1, CDC 7600, IBM 360/91, Manchester University MU5.
(iii) *Associative processors:* an example is the STARAN IV (Goodyear Aerospace).

Algorithms which exploit SIMD processors typically arise in connection with the following classes of problems:

(a) Matrix operations.
(b) Numerical integration of differential equations.
(c) Monte Carlo methods.
(d) Pattern recognition.

Array processors

This type is most frequently chosen as a model for an SIMD machine. Indeed, the schematic diagram in Fig. 2.2 of an SIMD machine corresponds to an array-processor. As shown, it consists of n processors which simultaneously execute the same instructions. The control unit analyses the program contained in the main store. All processors are driven centrally and the operation of a particular processor is determined not only by the sequence of instructions but also by the values calculated so far and the operands currently held by the processors. The idea of an array processor goes back to Zuse (cf. the drum-oriented Zuse K 58). The Illiac IV is a widely known array processor. An essential characteristic of this machine is that control of the SIMD structure is separated from other tasks and a general-purpose processor (the B 6500) undertakes all functions that are poorly suited to the SIMD structure. The SIMD part consists of four 'quadrants' which may be controlled separately or together. A quadrant is illustrated in Fig. 2.3.

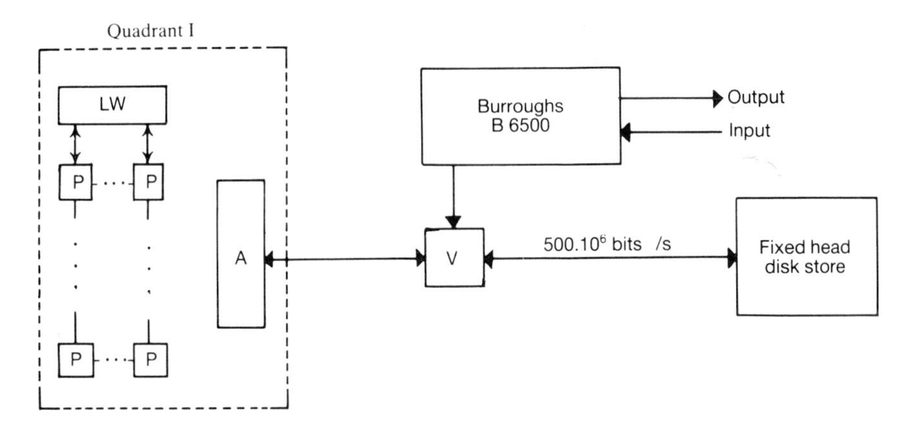

Fig. 2.3 – Quadrant of ILLIAC IV (P: processor with local store; A: exchange store; V: traffic control unit; LW: local control unit).

Each quadrant contains a matrix of 8 × 8 processors and a unit which manages local control. Each processor has an arithmetic unit with 64-bit word length and a module for the storage of local operands and of the program local to the quadrant. The local control unit takes its instructions from this store. The system is thus driven by interpreting these instructions, by extracting special words from the stores local to the processors, and by using operands of the neighbouring processors of the matrix.

Pipeline processors

Architecturally this type of computer is fundamentally different from the array processor. See Fig. 2.4. Instead of having several processors operating in parallel it possesses just one extremely elaborate processor, the pipeline. Thus, architecturally, it is not really a parallel processor at all. However, to execute operations which can be performed in parallel, these are broken down into elementary steps or stages. For example, floating point operations entail among others the steps of normalisation, exponent and mantissa calculation. Now, if an operation has been broken down into k stages it becomes possible to start executing the next overall operation before the previous one has been completed. Figure 2.5 illustrates this using a scalar vector multiplication as an example.

For n operations and k steps it can be shown [5] that the speed-up factor is given by

$$S = k - \frac{n}{k+n-1}$$

and thus, as $n \to \infty$, $S \sim k$ which means that, for large n, a pipeline processor behaves like an array processor with k processors.

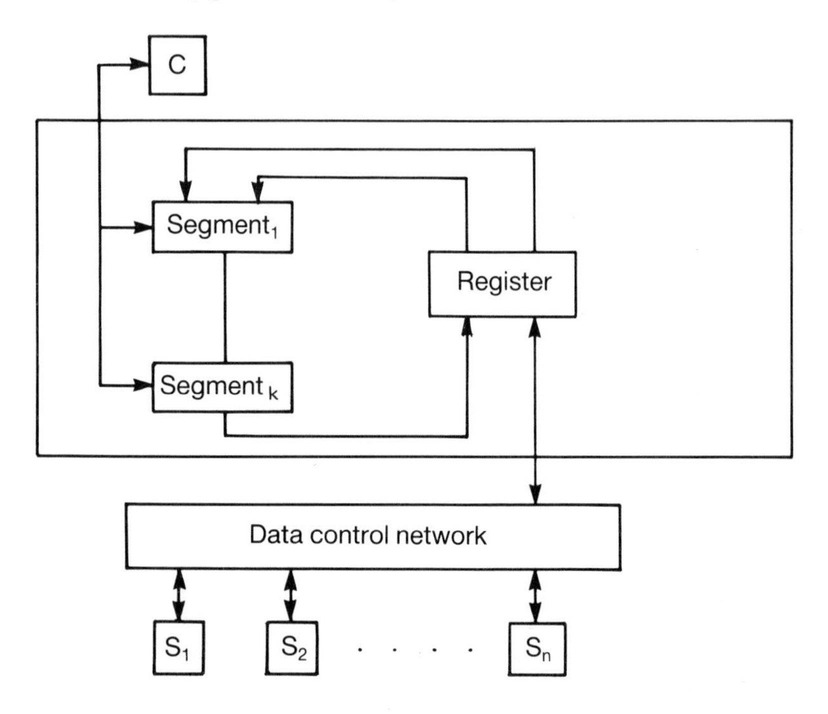

Fig. 2.4 – Model of a pipeline computer. (C: control unit; S: store).

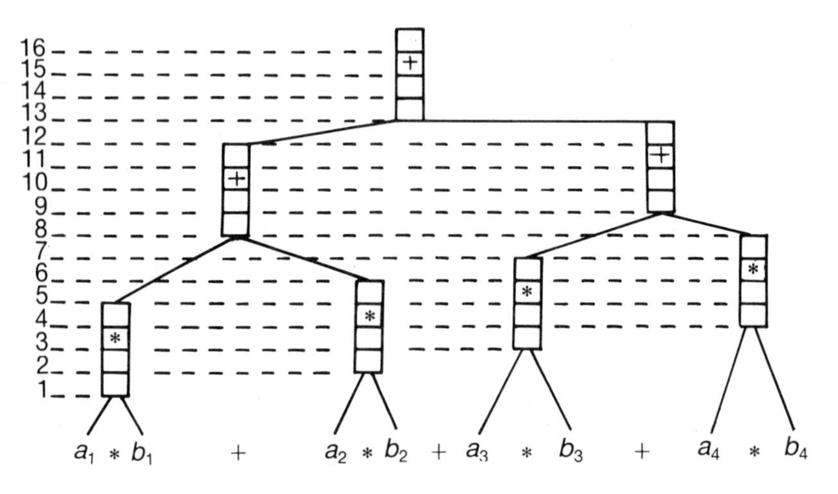

Fig. 2.5 – Illustrates the evolution in time of an inner product calculation on a pipeline computer.

Associative machines
In such machines an item in a storage location is retrieved by content rather than by address. This is naturally most sensible when the maximum number of processors in parallel interrogate the maximum number of storage areas. For example, the STARAN IV possesses 256 1-bit serial processors with one 256-bit store. Typical tasks for such machines are search procedures or pattern recognition.

2.2 MIMD machines

MIMD machines are less common than SIMD machines. MIMD structures are also quite different. Examples are the Korn multi-mini systems, the Carnegie–Mellon multisystem and, in the widest sense, some multiprocessor machines.

In contrast with the SIMD type an MIMD system possesses m independent processors $P_1, P_2, \ldots, P_m$ each having its own control unit. These processors share, among other things, a number of input–output units and a main store. At every instant the processor can carry out different instructions in parallel (refer to Fig. 2.2).

2.3 Remarks on associative machines and the Holland machine

The associative computer has a structure different from that of the classical computer. The reason is that, in addressing a storage element, it always at the same time tests (or processes) the content, and this means that the traditional distinction between store and processor is no longer clear or, indeed, always meaningful. The computer contains many elements with both properties. The SIMD principle, however, ensures that the control unit is always unequivocally separated from the rest of the machine. Since the MIMD principle entails that each processor has a control unit of its own it is not necessary to make a separate feature of the control unit. Carrying this line of thought further we can visualise a hypothetical machine in which there is no difference between the principal units – processor, store and control unit – and whose basic elements can perform all tasks on request. An example is provided by the Holland machine [6]. A cell-machine consists of a matrix of similar cells. Each cell contains a register for a few words and some simple switching networks. The transport routes for data and control signals follow the columns and rows of the matrix. The cells are to be considered to be endowed with a special function whenever commanded. The following diagram provides an example.

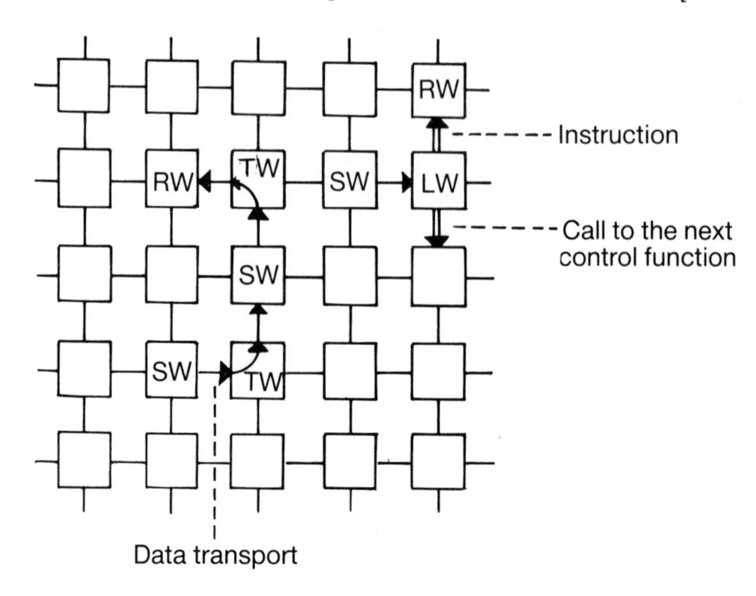

Fig. 2.6.

Each cell can, at a given moment, execute one of the following tasks:

1. Data flow (TW) — Transport of a datum or a call to a neighbouring cell.
2. Store (SW) — Passing on an operand, receiving an operand.
3. Computation (RW) — Linking of a stored operand with another operand. Result placed in the store allocated to the stored operand.
4. Control (LW) — Interpretation of a word in store as an instruction and progressing of calls to other cells (in particular as the last part of the operation: demand for the following control function which alerts the next operation).

2.4 Organisation of data

A big problem arises from the fact that the storage locations of a SIMD processor are in the end shared among the individual processors, while these have only limited communication with each other. It follows that *in algorithms for parallel processing the organisation and dynamic rearrangement of the data play a decisive role.*

As an example let us consider a 'mini-ILLIAC' with just three processors P_1, P_2, and P_3, each having three storage locations. Suppose that the elements of a 3×3 matrix $\mathbf{A} = [a_{ij}]$ are stored in their 'natural sequence' as shown in the diagram (2.3) below.

P_1	P_2	P_3
a_{11}	a_{12}	a_{13}
a_{21}	a_{22}	a_{23}
a_{31}	a_{32}	a_{33}

(2.3)

Now P_1 has access to a_{11}, a_{21} and a_{31}, and P_2 and P_3 to the elements shown in the columns below them, but P_1 does not have access to the elements in the second and third columns of $\mathbf{A}$. Parallel operation is thus possible on rows and diagonals, but not on columns. To store the elements in the form of the transpose $\mathbf{A}^T$ does not help. However, if the skew arrangement shown in diagram (2.4) is adopted the parallel processor can have access to the elements in rows, columns and diagonals. For instance, to have access in parallel to the elements in the second column we see that a_{12} is available to P_2, a_{22} to P_3 and a_{32} to P_1.

P_1	P_2	P_3
a_{11}	a_{12}	a_{13}
a_{23}	a_{21}	a_{22}
a_{32}	a_{33}	a_{31}

diagonal
1st column
3rd row

(2.4)

Some SIMD processors may have special data manipulation functions built in: the Fortran compilers for the CDC STAR-100 provide examples.

Comments on parallel storage

Parallel storage is particularly important in array processors, but can be equally significant for pipeline processors (see Fig. 2.4 and Appendix 1), although in this case other arrangements are often employed, for example the storage of vector operands only in a sequence of locations, one following the other, as in a list (STAR-100) or in locations with increasing addresses (ASC, CRAY-1). As we saw in the introductory example particular care must be exercised with storage for parallel processing to ensure parallel access. Further examples will now be considered.

Case 1

The computer uses m storage units and a one-dimensional vector with elements a_i is to be stored within these units.

1	2	3	4
a_1	a_2	a_3	a_4
a_5	a_6	a_7	a_8
a_9	a_{10}	a_{11}	a_{12}
.	.	.	.

(2.5)

The storage units are represented by the columns in the diagram. While the first m operands are being processed m further operands can be fetched, and so on. If, however, the array is so indexed that only odd-numbered elements can be called, the effective bandwidth is halved and there is obviously an access conflict. This can be avoided if m is a prime number for each index expressed in the scale of m can then be processed without conflict (see Definition (2.8)).

Case 2

Many programs contain multidimensional arrays. By analogy with the simple example of the skew storage arrangement of a 3 X 3 matrix let us try to find a scheme for an $n \times n$ matrix which, given m storage units, guarantees simultaneous access to rows. columns and diagonals.

As we saw in (2.3) the 'natural' arrangement permits access to rows and diagonals only, while the skew arrangement for $n = 3$ extended access also to columns. Consider now the array (2.6) where $n = 4$ processors each have $m = 4$ storage units.

1	2	3	4
(a_{11})	a_{12}	a_{13}	a_{14}
a_{24}	a_{21}	(a_{22})	a_{23}
(a_{33})	a_{34}	a_{31}	a_{32}
a_{42}	a_{43}	(a_{44})	a_{41}

(2.6)

The diagonal elements of the 'natural' ordering have been ringed and this makes it easy to see that access without conflict is possible to rows and to columns, but not to diagonals. It can be shown that for even n there is no way of arranging the elements of an $n \times n$ matrix in $m = n$ storage units so that arbitrary rows, columns and diagonals can be accessed without conflict. If, however, it is possible to use more than n storage units (columns) access without conflict to any row, column or diagonal can be had, and for other n-vectors too. In what follows we shall outline the notion of *skew storage.*

Let $m = 2^{2k} + 1$, where k is a positive integer. Then conflict-free access is available to rows, columns, diagonals and square blocks.

Example
$k = 1, \quad n = 4$

		Column				
		1	2	3	4	5
	1	a_{11}	a_{12}	a_{13}	a_{14}	
Row	2	a_{24}		a_{21}	a_{22}	a_{23}
	3	a_{32}	a_{33}	a_{34}		a_{31}
	4		a_{41}	a_{42}	a_{43}	a_{44}

Skew storage scheme:
$m = 5$ (2.7)

The 'natural' arrangement of this matrix $[a_{ij}]$ is

$$
\begin{matrix}
a_{11} & a_{12} & a_{13} & a_{14} \\
a_{21} & a_{22} & a_{23} & a_{24} \\
a_{31} & a_{32} & a_{33} & a_{34} \\
a_{41} & a_{42} & a_{43} & a_{44}
\end{matrix}
$$

and we now comment on its rearrangement.

The position (i', j') in the rearrangement of an element occupying position (i, j) in the natural arrangement, is defined by the transformation

$$i' = i, \quad j' = j + 2(i-1) \mod 5.$$

This may be expressed alternatively by stating that successive elements in the columns are stored $\delta_1 = 2 \pmod{m}$ storage units to the right of their predecessors, while successive elements in rows are stored $\delta_2 = 1$ storage units to the right of their predecessors. The whole is fixed by leaving a_{11} in its natural position. It is easily checked that elements of rows, columns and diagonals in the natural arrangment are relocated in different rows and columns of the skew storage scheme. The numbers δ_i are called displacement distances in the ith dimension and the whole scheme is a (δ_1, δ_2)-displacement (or skew) system (or scheme).

Thus, (2.7) is a (2,1) skew system with $m = 5$ while (2.6) is a (1,1) skew system with $m = 4$.

Generalisation to k dimensions is obviously possible in which case we refer to a $(\delta_1, \delta_2, \ldots, \delta_k)$-skew system for given m.

Definition (2.8)
An n-vector is called a d-ordered n vector modulo m if its ith element is stored in the μth storage unit, where

$$\mu = di + c \quad \mod m$$

and c is an arbitrary constant.

Example
We have already referred to this property with respect to (2.7). Specifically, row 1 is a 1-ordered 4-vector with $c = 0$. Row 2 is a 1-ordered vector with $c = 2$. Column 2 is a 2-ordered vector with $c = 0$. The main diagonal is a 3-ordered vector with $c = 3$.

Theorem (2.9)
A sufficient condition for conflict-free access to a d-ordered n-vector (mod m) is

$$m \geqslant nf \tag{2.10}$$

where f is the highest common factor of d and m.

Proof
The set S of storage units

$$S = \{\mu | \mu \equiv di + c \,(\mathrm{mod}\, m), \quad 0 \leqslant i \leqslant n-1\}$$

must contain n different elements. That is to say the storage elements in which the n elements of the vector are placed must be different. Suppose that S contains fewer than n elements. Then there exist integers i, j $(0, 1, 2, \ldots, n-1)$, $i \neq j$, such that

$$di + c \equiv dj + c \quad \mod m \;,$$

that is

$$d(i-j) \equiv 0 \quad \mod m \;,$$

which means that $d|i-j|$, which is different from zero, is divisible by m. If f is the highest common factor of m and d we may write $m = fm_1, d = fd_1$, where m_1 and d_1 are mutually prime. Thus $d_1|i-j|/m_1$ must be an integer, which means that m_1 must be a divisor of $|i-j|$. Hence

$$m_1 \leqslant |i-j| \leqslant n-1 < n,$$

and, accordingly,

$$m/f < n$$

in contradiction to (2.10). Q.E.D.

In a (δ_1, δ_2)-system it is now obvious that the columns are δ_1-ordered, the rows δ_2-ordered, and the diagonals $(\delta_1 + \delta_2)$-ordered vectors. If it is required to have access without conflict to these three types of n-vector Theorem (2.9) requires the following three conditions to be satisfied:

$$
\begin{aligned}
& m \geqslant n\ (\delta_1, m) && \text{for columns;} \\
& m \geqslant n\ (\delta_2, m) && \text{for rows;} \\
& m \geqslant n\ (\delta_1 + \delta_2, m) && \text{for diagonals;}
\end{aligned}
\tag{2.11}
$$

where (a, b) has its usual number-theoretic meaning as the highest common factor of a and b.

3

Fundamentals of parallel numerical analysis

3.1 Complexity

In this section we shall take a brief look at questions of complexity in the context of numerical methodology. In the sense of Chapter 1 time and processor bounds are often sought for a given algorithm A. Starting with the speed-up ratio $S_n(A)$ for the algorithm in question, defined in (1.2) by

$$S_n(A):= T_1/T_n$$

we introduce the following additional notation:

$N(A)$: = the least number of processors required to obtain a maximum speed-up for the given algorithm A;

$T_N(A)$: = number of time units required when using $N(A)$ processors;

$T_p(A)$: = number of time units required when the number of processors available is restricted to p ($< N(A)$).

We shall examine the important and typical case of time and processor bounds in the evaluation of arithmetical expressions and the discussion will be limited to transformations of an expression that exploit the associative, commutative and distributive properties. These transformations enable parallel computers to achieve asymptotic speed-up factors of the order $0(n/\log_2 n)$, and effectiveness ratios of the order $0(1/\log_2 n)$, where n is the number of operands.

Definition (3.1)

An arithmetic expression A is a well-defined string consisting of the four arithmetic operations $(+, -, *, /)$, left- and right-hand brackets, and *atoms* which are the constants and variable operands. An arithmetic expression containing n different atoms will be denoted by $A\langle n\rangle$.

Using a single processor we know that the time required to evaluate $A\langle n\rangle$ is $n-1$ time units. With an arbitrary number of processors $A\langle n\rangle$ can, in favourable circumstances, be evaluated in $\log_2 n$ time units.

Example: Expression with 8 atoms

$$A\langle 8\rangle = \sum_{0}^{8} a_i \quad .$$

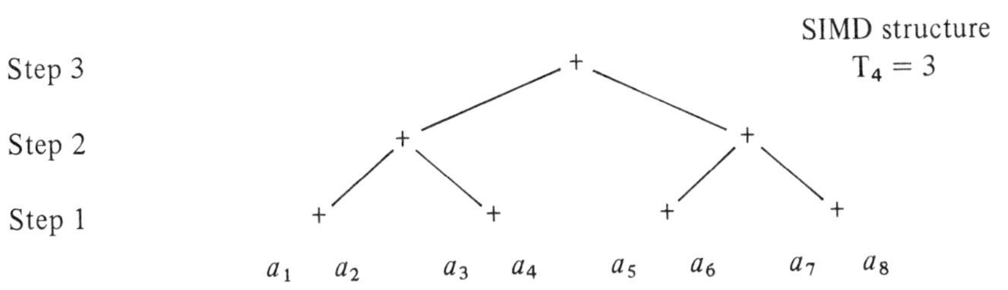

Here $\mathbf{a} = (a_1, a_2, \ldots, a_8)$ would be stored as a vector with 8 elements (an array) in the computer. Let us denote by $\lceil x \rceil$ the smallest integer greater than or equal to x. The following lemma provides a bound on the time reqired to compute an arbitrary expression.

Lemma (3.2)
For an arbitrary arithmetic expression $A\langle n\rangle$ we have

$$T(A\langle n\rangle) \geqslant \lceil \log_2(n) \rceil \quad .$$

On the other hand, expressions can be constructed whose computation requires $0(n)$ time units independently of how many processors are available. Consider, for example, the Horner scheme for the evaluation of a polynomial of degree n:

$$P_n(x) = \sum_{0}^{n} a_i x^i = a_0 + x(a_1 + x(a_2 + \ldots + x(a_{n-1} + xa_n) \ldots) \quad .(3.3)$$

$0(n)$ operations can not be avoided since a strict sequential ordering is fixed by the brackets. The availability of more than a single processor has therefore no power to speed up the calculation. However, we are not constrained to keep arithmetic expressions in the form given. By using associativity, commutativity and distributivity we may be able to transform a given expression A into an equivalent form better, specifically more quickly, capable of evaluation.

To this end consider the following simple expressions:

Example 1

$$A\langle 4\rangle := (((a + b) + c) + d), \; a, b, c, d \in \mathbb{R} \quad . \tag{3.4}$$

The analysis of this expression will be effected by the use of *parse trees*. The straightforward parse tree for (3.4) involves three steps:

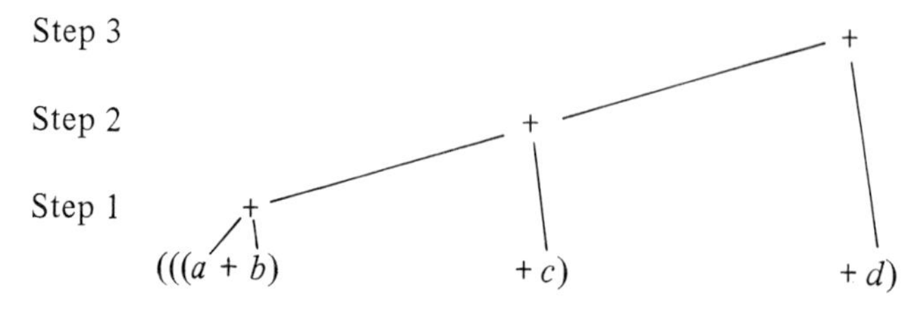

with $T_1 = 3$.

This implementation possesses a tree height of three steps. If we use the associative property of addition the brackets can be placed differently:

$$A\langle 4\rangle = (a+b)+(c+d) \ . \qquad (3.5)$$

This expression can be dealt with using a tree height of only two and implied SIMD structure.

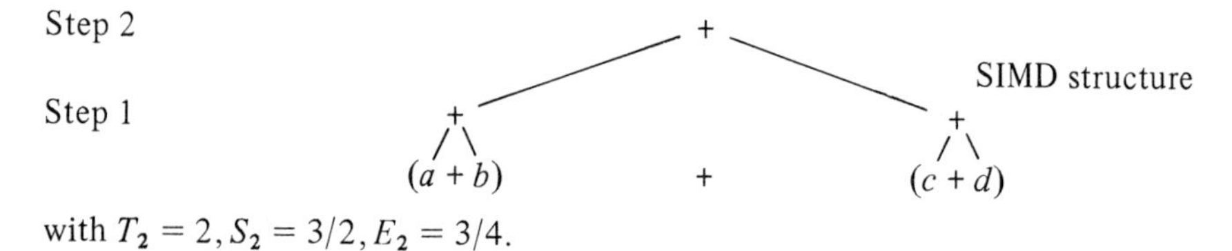

with $T_2 = 2, S_2 = 3/2, E_2 = 3/4$.

Note that in both cases three additions have to be carried out.

Example 2

Let the expression be

$$A\langle 4\rangle := a + bc + d, \quad a, b, c, d \in \mathbb{R} \ . \qquad (3.6)$$

The parse tree is shown below. Its height is three. It is not a unique tree, but no lower level tree can be obtained by exploitation of associativity.

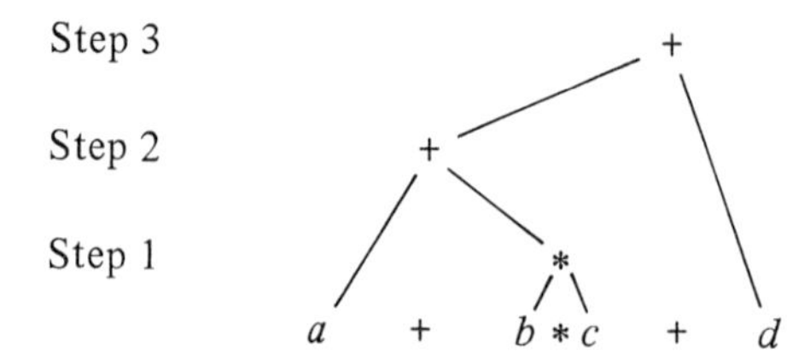

The time required $T_1 = 3$.

However, by using the commutative property of addition we can also write

$$A\,\langle 4\rangle = a + d + bc, \tag{3.7}$$

and now a tree height of two is obtainable and implied MIMD structure.

```
Step 2                     +                    MIMD structure
                        /     \
Step 1                +         *
                     / \       / \
                    a + d   +  b * c
```

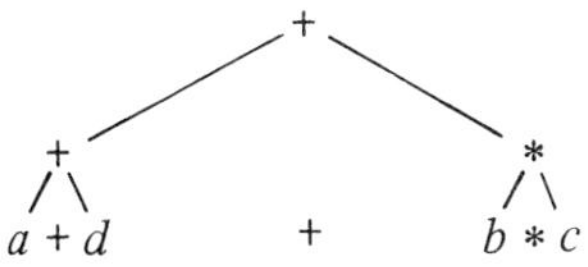

with $T_2 = 2, S_2 = 3/2, E_2 = 3/4$.

It is natural to ask the following questions.

(1) How many tree height reductions can be achieved for a given arithmetic expression?
(2) Can general bounds and algorithms be given for tree height reduction?
(3) How many processors are needed for optimality, in some sense?

Other questions come readily to mind.

Beatty, Baer, Bovet and Kuck have investigated these matters. Kuck and Muraoka [7] gave an upper bound for tree height reduction on the assumption that only associativity and commutativity are used.

Theorem (3.8)
Let $A\,\langle n|d\rangle$ be an arithmetic expression with a depth d of nested brackets. The properties of associativity and commutativity allow $A\,\langle n|d\rangle$ so to be transformed that

$$T_p(A\,\langle n|d\rangle) \leqslant \lceil \log_2 n\rceil + 2d + 1, \tag{3.9}$$

where $p \leqslant \lceil n/2 - d\rceil$.

3.2 Principles for the construction of parallel algorithms

From the above discussion we can discern that the kind of parallelism plays a central role in parallel data processing, particularly where numerical algorithms are concerned, for example in the arrangement and dynamical sorting of data needed by the parallel algorithm during the program.

A *first principle* in the construction of a parallel algorithm for most SIMD machines is to start with a serial algorithm and to convert it into a procedure which operates on vectors, the reason being that vector operations can be carried out in parallel. This principle will be explained by the example of the solution of a $n \times n$ system of linear equations with a lower triangular coefficient matrix.

$$\mathbf{A\,x} = \mathbf{b} \leftrightarrow \begin{bmatrix} a_{11} & & 0 & \\ a_{21} & a_{22} & & 0 \\ \vdots & & \ddots & 0 \\ a_{n1} & & & a_{nn} \end{bmatrix} \begin{bmatrix} x_1 \\ x_2 \\ \vdots \\ x_n \end{bmatrix} = \begin{bmatrix} b_1 \\ b_2 \\ \vdots \\ b_n \end{bmatrix}, \quad (3.10)$$

where **A** is nonsingular. Using a serial computer n^2 arithmetic operations are necessary to effect the inversion. It can be shown, writing

$$\mathbf{L}^{(i)} := \begin{bmatrix} 1 & & & & 0 \\ & \ddots & & & \\ & & 1/a_{ii} & & \\ 0 & & \vdots & \ddots & \\ & & -a_{ni}/a_{ii} & \cdots & 1 \end{bmatrix}, \quad i = 1, 2, \ldots, n,$$

that

$$\mathbf{x} = \mathbf{L}^{(n)}\,\mathbf{L}^{(n-1)} \ldots \mathbf{L}^{(1)}\,\mathbf{b}\;. \quad (3.11)$$

The following vectors $\mathbf{y}^{(1)}, \mathbf{y}^{(2)}, \ldots, \mathbf{y}^{(n+1)}$ must be calculated.

$$\mathbf{y}^{(i+1)} := \mathbf{L}^{(i)}\,\mathbf{y}^{(i)} = \begin{bmatrix} y_1^{(i)} \\ \vdots \\ y_i^{(i)}/a_{ii} \\ y_{i+1}^{(i)} - a_{i+1,i}\,y_i^{(i)}/a_{ii} \\ \vdots \\ y_n^{(i)} - a_{ni}\,y_i^{(i)}/a_{ii} \end{bmatrix}, \quad i = 1, 2, \ldots, n; \quad (3.12)$$

with $\mathbf{y}^{(1)} := \mathbf{b}$.

If n processors are available the solution **x** can be calculated in about $3n$ steps, that is $0(n)$. The speed-up is $S_n = n/3$ and the efficiency $E_n = 1/3$. If, however, merely $k < n$ processors are available then, corresponding to n serial steps, $\lceil n/k \rceil$ parallel steps are necessary. For the calculation of (3.12), therefore, $0(n^2/k)$ steps are necessary, giving a speed-up of $S_k = 0(k)$ and efficiency $E_k = 0(1)$.

Example

It may help the reader to see the details of a concrete example. The serial procedure is equivalent to setting out the array $\mathbf{A}|\mathbf{b}$ and systematically reducing it to $\mathbf{I}|\mathbf{A}^{-1}\mathbf{b}$ by division of rows by diagonal terms and reduction to zero of subdiagonal terms in columns by multiplication and subtraction. Thus to deal with the first row requires one division (through the first row by a_{11}). To reduce the (2,1) place to zero involves replacing the second element of the $\mathbf{b}$ vector by $b_2 - a_{21} * (b_1/a_{11})$, that is a multiplication and subtraction. The processing of the first column thus involves $1 + 2(n-1)$ operations. Then for the second column and subdiagonal terms $1 + 2(n-2)$ operations are necessary. The whole process requires thus n divisions plus $2(1 + 2 + 3 + \ldots + n-1)$ multiplications and subtractions, a total of $n + 2 \,.\, \frac{1}{2}(n-1)n = n^2$ arithmetical operations in sequence.

For n=4 the proposed procedure is as follows:

$$\mathbf{L}_1 = \begin{bmatrix} -\dfrac{1}{a_{11}} & 0 & 0 & 0 \\ -\dfrac{a_{21}}{a_{11}} & 1 & 0 & 0 \\ -\dfrac{a_{31}}{a_{11}} & 0 & 1 & 0 \\ -\dfrac{a_{41}}{a_{11}} & 0 & 0 & 1 \end{bmatrix} ; \mathbf{y}_1 = \begin{bmatrix} b_1 \\ b_2 \\ b_3 \\ b_4 \end{bmatrix} \text{ and } \mathbf{y}_2 = \mathbf{L}_1\mathbf{y}_1 = \begin{bmatrix} \dfrac{b_1}{a_{11}} \\ \dfrac{-a_{21}b_1}{a_{11}} + b_2 \\ \dfrac{-a_{31}b_1}{a_{11}} + b_3 \\ \dfrac{-a_{41}b_1}{a_{11}} + b_4 \end{bmatrix}$$

This is the same as the procedure to deal with the first row and column of the array described above but with four processors the operation could be effected in parallel as shown below. (In these diagrams the symbols at the roots of the trees are the result of the operations illustrated.)

Processor

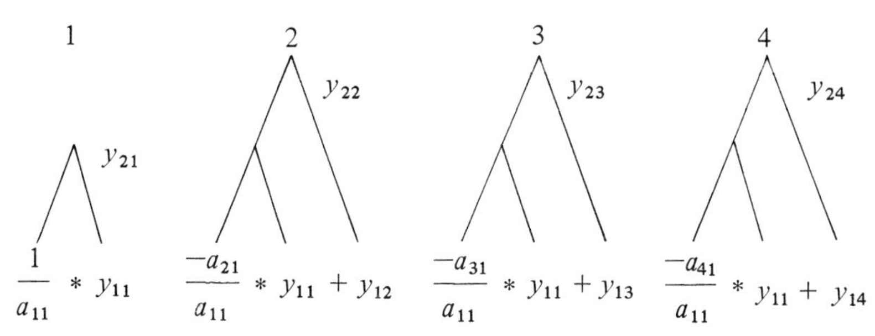

The time required to execute all operations in parallel corresponds to division (by a_{11}), multiplication and addition: 3 time units. Note that $x_1 = y_{21}$. The next step uses

$$\mathbf{L}_2 = \begin{bmatrix} 1 & 0 & 0 & 0 \\ 0 & \dfrac{1}{a_{22}} & 0 & 0 \\ 0 & \dfrac{-a_{32}}{a_{22}} & 1 & 0 \\ 0 & \dfrac{-a_{42}}{a_{22}} & 0 & 1 \end{bmatrix}$$

to operate on $\mathbf{y}_2 = (y_{21}, y_{22}, y_{23}, y_{24})$ to get $\mathbf{y}_3$. Since $x_1 = y_{21}$ only three components, y_{32}, y_{33} and y_{34} are needed in the next step and only three processors need be used.

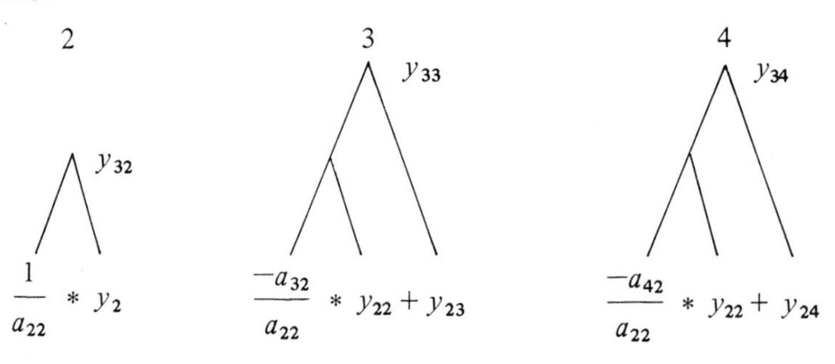

Again 3 time units have been used and $x_2 = y_{32}$. The remaining two steps can be shown schematically.

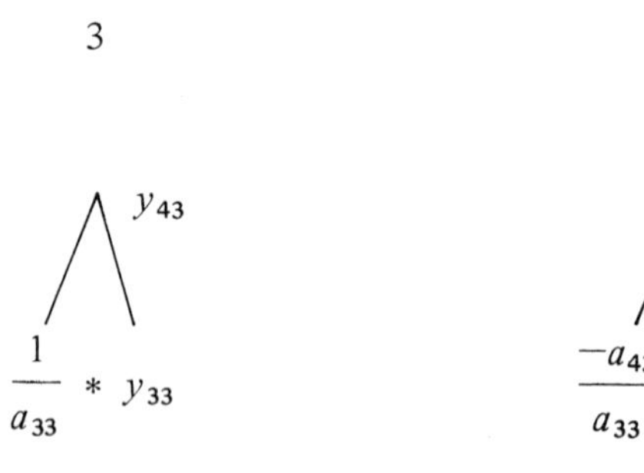

This gives $x_3 = y_{43}$. Finally

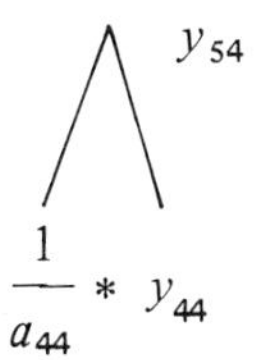

$x_4 = y_{54}$. All steps have used 3 time units except the last which requires only 2. Thus for general n the time steps are $3n-1$.

A second principle is the method of *vector iteration.* This entails substituting an iterative parallel algorithm for a direct (non-iterative) serial algorithm. The speed-up factor of the parallel version depends on the ratio of the steps needed in the direct version to those required by the iteration. We shall now explain the process by using an example of Heller [8] for triangular decomposition of a tridiagonal matrix **A**.

$$\mathbf{A} = \begin{bmatrix} d_1 & f_1 & & & 0 \\ e_2 & \ddots & \ddots & & \\ & \ddots & \ddots & \ddots & \\ & & \ddots & \ddots & f_{n-1} \\ 0 & & & e_n & d_n \end{bmatrix} = \mathbf{L}\ \mathbf{U} \tag{3.13}$$

where

$$\mathbf{L} = \begin{bmatrix} 1 & & 0 \\ m_2 & \ddots & \\ 0 & m_n & 1 \end{bmatrix}, \quad \mathbf{U} = \begin{bmatrix} u_1 & f_1 & 0 \\ & \ddots & f_{n-1} \\ 0 & & u_n \end{bmatrix} \tag{3.14}$$

The u_i are calculated from the scheme

$$u_1 := d_1, \quad u_i := d_i - e_i\, f_{i-1}/u_{i-1}, \quad 2 \leqslant i \leqslant n. \tag{3.15}$$

The m_i can then be calculated in parallel:

$$m_i := e_i/u_{i-1}\ , \qquad i = 2, \ldots, n\ .$$

(3.15) is a serial procedure which can be parallelised by the iteration:

$$\begin{aligned} u_i^{(0)} &= d_i \\ u_i^{(j)} &:= d_i - e_i f_{i-1}/u_{i-1}^{(j-1)} \quad , \quad i = 1, 2, \ldots, n, \end{aligned} \tag{3.16}$$

where $u^{(j)}$ is the jth iterate.

This parallelisation can be a reasonable one if the computer can carry out operations with vectors of n components essentially faster than n scalar operations. Moreover the number of iterations required must naturally be significantly less than n.

A *third* and very general principle is the method of recursive doubling due to Kogge [9] to which we have already referred in discussing complexity questions. To describe this method consider a set $M: = (m_1, m_2, \ldots, m_N)$ having $N = 2^n$ elements. Let **op** be an arbitrary associative operation that can be carried out on M, for example $\mathbf{op} \in (+, *, \max, \ldots)$. Now the expression $m_1 \,\mathbf{op}\, m_2 \,\mathbf{op} \ldots \mathbf{op}\, m_N$ will be formed both serially and in parallel and a comparision will be made. For $N = 4$ we have the following scheme:

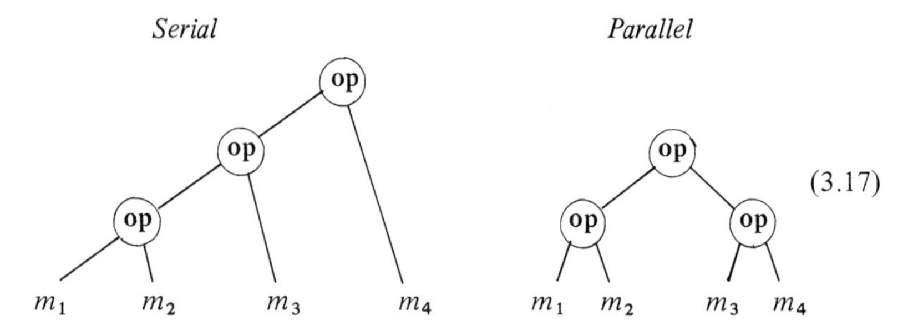

(3.17)

Generally, recursive doubling with $N = 2^n$ elements requires $\log_2 N$ parallel steps. Serial implementation requires $N-1$ steps.

As an example we now consider a second order recurrence formula. This is essentially a serial process (see also section 3.3).

$$\begin{aligned} f_0 &:= 1, \qquad f_1 := b, \\ f_i &:= b_i f_{i-1} + a_i f_{i-2}, \; a_i, \; b_1, b_i \in \mathbb{R} \; , \; 2 \leqslant i \leqslant N = 2^n . \end{aligned} \tag{3.18}$$

This rule yields a sort of Fibonacci sequence.

Let

$$\mathbf{F}_j = \begin{bmatrix} f_j \\ f_{j-1} \end{bmatrix} \text{ and } \mathbf{A}_i := \begin{bmatrix} b_i & a_i \\ 1 & 0 \end{bmatrix}, \quad (j=1, \ldots, N; \; i=2, \ldots, N). \tag{3.19}$$

Then (3.18) can be written as

$$\mathbf{F}_i = \mathbf{A}_i \mathbf{F}_{i-1} = \mathbf{A}_i \mathbf{A}_{i-1} \ldots \ldots \mathbf{A}_2 \mathbf{F}_1 \tag{3.20}$$

$$\mathbf{F}_N = \mathbf{A}_N \ldots \mathbf{A}_2 \mathbf{F}_1 \ . \tag{3.21}$$

Since matrix vector multiplication is associative $\mathbf{F}_N$ can be computed using the recursive doubling technique in $0(\log_2 N)$ steps by comparison with $0(N)$ steps using a serial calculation.

This general principle will be illustrated by further consideration of the LU decomposition of a tridiagonal matrix $\mathbf{A}$ (see (3.13) – (3.15)). The m_i can be calculated in parallel when the u_i are known. To determine u_i from (3.15) we define

$$q_0 := 1, \ q_1 := d_1, \ q_i := u_i q_{i-1}, \ 2 \leqslant i \leqslant n \ .$$

Then

$$u_i = q_i / q_{i-1} \ . \tag{3.22}$$

Hence it follows from (3.15) that

$$q_i := d_i q_{i-1} - e_i f_{i-1} q_{i-2}, \ 2 \leqslant i \leqslant n \ . \tag{3.23}$$

Then by defining

$$\mathbf{Q}_j := \begin{bmatrix} q_j \\ q_{j-1} \end{bmatrix}, \quad \mathbf{D}_i := \begin{bmatrix} d_i & -e_i f_{i-1} \\ 1 & 0 \end{bmatrix}$$

for $1 \leqslant j \leqslant n, \ 2 \leqslant i \leqslant n$,
we get

$$\mathbf{Q}_i := \mathbf{D}_i \mathbf{D}_{i-1} \ldots \mathbf{D}_2 \mathbf{Q}_1 \ , \quad 2 \leqslant i \leqslant n. \tag{3.24}$$

Thus $\mathbf{Q}_n$ can be calculated by the recursive doubling technique in $0(\log_2 n)$ steps and simultaneously we obtain the products $Q_i, i=2,3, \ldots, n$. The u_i are obtained from (3.22) and can also be calculated in $0(\log_2 n)$ steps. In the serial mode $0(n)$ steps would be needed. Thus we have a speed-up $S_n = 0(\log_2 n/n)$, typical of the recursive doubling technique.

Remarks on the numerical quality of parallel algorithms

Stability, rounding errors, and the propagation of errors in relation to parallel algorithms, have not so far been investigated extensively. Moreover, it happens in many cases that parallel processing leads to numerically inferior results, but this is not always so. We shall show by means of an example how the parallel version of an algorithm actually leads to better results than the serial.

Consider the sum

$$S_N = \sum_{k=1}^{N=2^n} a_k \quad . \tag{3.25}$$

(a) *Serially*

$$S_0 := 0, \; S_k := S_{k-1} + a_k \;, \quad k = 1, 2, \ldots, N.$$

Because of rounding errors we actually deal with approximate values S'_k and a'_k. Thus

$$S'_0 = 0, \; S'_k = S'_{k-1} + a_k + r_k \tag{3.26}$$

where

$$r_k = a'_k - a_k \quad .$$

Let

$$e_k := S'_k - S_k = \sum_{j=1}^{k} r_j \quad .$$

If the mantissa in the computer has s binary places then S'_k will be expressed in the form

$$|S'_k| = 0 . b_1 \, b_2 \, b_3 \ldots b_s \, 2^{p(k)} \quad . \tag{3.27}$$

If a is defined by

$$a := \max_k \; |a_k|$$

then

$$|S_k| \leqslant ka \;,$$

and

$$2^{p(k)} \leqslant 2ka \;.$$

Hence

$$|e_N| = |S'_N - S_N| \leqslant \sum_{j=1}^{N} |r_j| < \sum_{j=1}^{N} 2^{-s} \, 2^{p(j)}$$

$$\leqslant 2^{-s+1}\, a \sum_{j=1}^{N} j \;=\; 2^{-s} a\, N(N+1)\ .$$

Thus the error is $0(N^2)$.

(b) *Parallel mode*

Let

$$S_{0i} := a_i\ ,\quad i = 1, 2, \ldots, N,$$

$$S_{ki} := S_{k-1,2i-1} + S_{k-1,2i}\ , \qquad k=1,2,\ldots,\log_2 N$$

$$i=1,2,\ldots,N/2^k\ .$$

Proceeding as in (a) above we get

$$|S_{ki}| \leqslant 2^k a, \quad 2^{p(k)} \leqslant 2{\cdot}2^k a\ .$$

Then for the error e_{ki} arising at each calculation of S_{ki} we have

$$|e_{ki}| \leqslant 2^{-s}\, 2{\cdot}2^k a\ .$$

For the rth ($r = N/2^k$) partial sum of the kth step we obtain for the estimate of the error

$$e_k := \sum_{i=1}^{r} e_{ki}\ ,$$

the bound

$$|e_k| \leqslant \frac{N}{2^k} \cdot 2^{-s} \cdot 2 \cdot 2^k a \;=\; N a\, 2^{-s+1}\ .$$

Hence, the estimate of the overall error

$$e := \sum_{k=1}^{\log_2 N} e_k$$

satisfies

$$|e| \leqslant \sum_{k=1}^{\log_2 N} N a\, 2^{-s+1} = 2^{-s+1} N a \log_2 N = 0(N \log_2 N), \qquad (3.28)$$

an improvement on the serial result.

3.3 Recurrence relations

3.3.1 Introduction

In numerical analysis the solution of a problem is often expressed as a sequence $x_1, x_2, \ldots, x_n, x_i \in \mathbb{R}$, where each x_i $(i=1, 2, \ldots, n)$ may depend on the x_j with smaller index $(j<i)$. Equations possessing this property are called recurrence equations. Their solution for given boundary conditions constitutes a recurrence problem [10].

As a general example of a problem of this kind we shall use the representation of a time-dependent linear system in which the state x_t of the system at time t is expressed as a linear function of the state at time $t-1$. Then

$$x_1 = c_1 \; ,$$

the boundary condition, together with

$$x_2 = a_2 \, x_1 + c_2$$

$$\vdots$$

$$x_t = a_t x_{t-1} + c_t \quad \text{(the general recurrence equation)}$$

$$\vdots$$

$$x_n = a_n \, x_{n-1} + c_n \; .$$

At first sight the problem may look highly serial and particularly suited to the conventional SISD computer. Linear recurrence relations are of central importance in computer design and in the use of computers. Whereas for instance, an arithmetic expression defines a fixed calculation procedure with a scalar result, a recurrence equation is a dynamic procedure for the computation of a scalar or of arrays of results. In the computer applications mentioned above linear recurrences are found in software and systems analysis. It is thus a self-evident task to develop rapid and efficient algorithms for their solution. We begin with some simple examples.

1. The inner product of two vectors

   $$\mathbf{x} = (x_1, x_2, \ldots, x_n) \text{ and } \mathbf{y} = (y_1, y_2, \ldots, y_n)$$

   can be written as a linear recurrence equation of the first order

   $$z := z + x_k y_k, \quad (1 \leqslant k \leqslant n)$$

   with $z := 0$ as boundary condition.

2. The Horner scheme for the evaluation of a polynomial $P_n(x) = \sum_1^n a_i x^i$ for

a special value $x = x_0$ of the argument, is a linear recurrence with a scalar value:

$$p := a_{n-k} + x_0 p, \quad (1 \leqslant k \leqslant n) \tag{3.30}$$

with

$$p := a_n$$

at the start of the calculation. After n iterations one obtains for (3.29) and (3.30) expressions which can be represented on parse trees with $\log_2 n$ levels: that means that they can be calculated in $0(\log_2 n)$ time steps when $N=0(n)$ processors are available.

3. The second order Fibonacci sequence

$$f_k := f_{k-1} + f_{k-2}, \quad (3 \leqslant k \leqslant n), \tag{3.31}$$

$$(f_1 = f_2 = 1)$$

can be regarded as a second order recurrence equation which generates a vector result in two dimensions.

3.3.2 Linear recurrence relations and algorithms for their solution

Definition (3.32)

A linear recurrence system $R\langle n,m\rangle$ of order m for n equations is defined for $m \leqslant n-1$ by

$$R\langle n,m\rangle : x_k := \begin{cases} 0 & (k \leqslant 0) \\ c_k + \sum_{j=k-m}^{k-1} a_{kj} x_j & (1 \leqslant k \leqslant n)\ . \end{cases} \tag{3.33}$$

When $m = n-1$ this system is called an *ordinary linear system of recurrence equations* and denoted by $R\langle n\rangle$.

Using matrix-vector notation we can write (3.33) as

$$\mathbf{x} = \mathbf{c} + \mathbf{A}\mathbf{x} \tag{3.34}$$

where

$$\mathbf{x} = (x_1, \ldots, x_n)\,, \quad \mathbf{c} = (c_1, \ldots, c_n),$$

and

$$\mathbf{A} = [a_{ik}\colon i,\ k = 1, 2, \ldots, n]$$

is a strictly lower triangular matrix with $a_{ik} = 0$ for $i \leqslant k$ or $i-k > m$. **A** is a band matrix for $m < n-1$.

Example

1. $n = 4,\ m = 2.$ The system is $R\ \langle 4, 2\rangle$:

$$
\begin{aligned}
x_1 &= c_1 \\
x_2 &= c_2 + a_{21}\, x_1 \\
x_3 &= c_3 + a_{31}\, x_1 + a_{32}\, x_2 \\
x_4 &= c_4 \qquad\qquad + a_{42}\, x_2 + a_{43}\, x_3 \ .
\end{aligned}
$$

2. For $n = 5$ we have the following ordinary linear recurrence $R\ \langle 5\rangle$:

$$
\begin{aligned}
x_1 &= c_1 \\
x_2 &= c_2 + a_{21}\, x_1 \\
x_3 &= c_3 + a_{31}\, x_1 + a_{32}\, x_2 \\
x_4 &= c_4 + a_{41}\, x_1 + a_{42}\, x_2 + a_{43}\, x_3 \\
x_5 &= c_5 + a_{51}\, x_1 + a_{52}\, x_2 + a_{53}\, x_3 + a_{54}\, x_4 \ .
\end{aligned}
$$

Before embarking on special algorithms for the solution of these systems we must discuss a basic concept of many parallel algorithms used for recurrences. This is the so-called *log-sum algorithm*. It is based on the *recursive doubling* principle (see Stone [11]) and is especially suitable for a pure SIMD computer. A classic example for the log-sum algorithm is the simple first order linear recurrence:

$$x_k := a_k + x_{k-1} \ , \qquad 1 \leqslant k \leqslant n \ ,$$

with x_0 a given initial value.

To calculate x_k in parallel mode it is merely necessary to recall (3.4) and (3.5) where the associative law of addition was used. We have from the recurrence

$$x_4 := a_4 + (a_3 + (a_2 + a_1)),$$

which may be written instead

$$x_4 := (a_4 + a_3) + (a_2 + a_1) \ .$$

For $n = 8$ we have $x_8 = \sum_1^8 a_k$, and the following diagram recalls the now familiar steps and timings

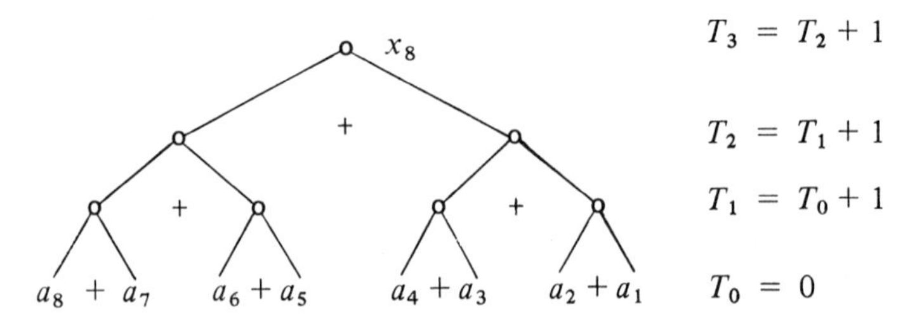

With $n = 2^N$ elements this procedure requires only $\log_2 n$ parallel time steps instead of $n-1$ in serial mode.

Generally this method depends on the decomposition of an expression A_0 into two simpler expressions A_1, A_2, each of which can be calculated simultaneously on its own processor. To achieve this the following conditions must be satisfied:

(a) A function f must exist such that $A_0 = f(A_1, A_2)$
(b) The calculation of A_1 and A_2 are independent processes and A_1 and A_2 are of equal computational complexity.
(c) A_1 and A_2 require the same sequence of operations for their calculation.

By further decomposition of A_1 and A_2 according to these principles we arrive finally by the principle of recursive doubling at A_0. Recall also the tree height reduction of arithmetic expressions. Many recurrence relations are not associative and are therefore not capable of resolution by this fast and direct method. (See *companion-functions.*) The principle is found to be applicable in a number of different parallel algorithms.

Another procedure leads to the *column-sweep* algorithm. This solves an $R\langle n\rangle$ system by an acceptably fast and efficient method such that with $N = 0(n)$ processors the system can be computed in $0(n)$ time unit steps.

Column-Sweep Algorithm

Step 1: We refer back to the set of equations at the beginning of this example. x_1 is known. Then the expressions $a_{i1}\, x_1 + c_i$ can be calculated in parallel. Now x_2 is known and there remains an $R\langle n-1\rangle$ system.

Step 2: Now that x_2 has been calculated the expressions $a_{i2}\, x_2 + c_i^{(1)}$ with $c_i^{(1)} := a_{i1}\, x_1 + c_i$ are calculated in parallel, and then x_3 is known. This procedure continues and in general . . .

Step k: x_k is known and the expressions $a_{ik} x_k + c_i^{(k-1)}$ are calculated where $c_i^{(k-1)} := a_{i,\,k-1}\, x_{k-1} + c_i^{(k-2)}$, giving x_{k+1}. Continuing thus we arrive at . . .

Step n–1: The calculation of x_n completes the calculation of the vector $\mathbf{x}$. This simple algorithm requires $n-1$ processors at step 1 and fewer thereafter, namely $T_p = 2(n-1)$, with $p = n-1$.

The number of time unit steps for the solution of $R\langle n\rangle$ with a single processor is

$$T_1 = n(n-1) \ .$$

Thus, for the speed-up factor we have

$$S_p = T_1/T_p = n/2 \ ,$$

and for the efficiency

$$E_p = S_p/p = n/\{2(n-1)\} > ½ \ .$$

Obviously the column-sweep algorithm for an $R\langle n,m\rangle$ system has speed-up factor $S_m = 0(m)$.

In applications, $R\langle n,m\rangle$ systems with $m \ll n$ are of interest. Frequently m is not more than 2. However, m is the maximal number of processors used and when a large number of processors is available the procedure is disadvantageous.

One of the fastest procedures for the resolution of an $R\langle n,m\rangle$ system having small m is the so-called *recurrent product form algorithm.* The following theorem is relevant [12].

Theorem

Every $R\langle n,m\rangle$ system can be evaluated in T_p time steps where

$$T_p < (2 + \log_2 m) \log_2 n - \tfrac{1}{2} \log_2 m (\log_2 m + 1)$$

and

$$p \leqslant \begin{cases} \tfrac{1}{2}m(m+1)n + 0(m^3) & \text{for } 1 \leqslant m \leqslant \tfrac{1}{2}n \\ \dfrac{n^3}{68} + 0(n^2) & \text{for } \tfrac{1}{2}n < m \leqslant n-1 \,. \end{cases}$$

In the particular case of an $R\langle n,2\rangle$ system with large n the necessary time and number of processors for the evaluation of the system can be shown by an exact calculation to satisfy

$$T_p \leqslant (2 + \log_2 2) \log_2 n - \tfrac{1}{2} \log_2 2(\log_2 2+1) = 3 \log_2 n - 1,$$

with

$$p \leqslant 3n - 8$$

in accordance with the Theorem.

The recurrent product form algorithm is developed from the product form representation of the solution of the $R\langle n\rangle$ system

$$\mathbf{x} = \mathbf{A}\,\mathbf{x} + \mathbf{c} \tag{3.35}$$

where $\mathbf{A}$ is a strict lower triangular matrix. The solution of (3.35) is

$$\mathbf{x} = (\mathbf{I}-\mathbf{A})^{-1}\,\mathbf{c} = \mathbf{L}^{-1}\,\mathbf{c}\ , \tag{3.36}$$

say. As effectively shown in section 3.2 and [9], $\mathbf{L}^{-1}$ can be expressed in the product form

$$\mathbf{L}^{-1} = \mathbf{M}_n\,\mathbf{M}_{n-1} \ldots \mathbf{M}_1 \tag{3.37}$$

with

$$\mathbf{M}_i = \begin{bmatrix} 1 & & & & & 0 \\ & \ddots & & & & \\ & & 1 & & & \\ & & & 1/a_{ii} & & \\ & & & \vdots & 1 & \\ & & & -a_{i+1\,i}/a_{ii} & & \ddots \\ 0 & & & -a_{ni}/a_{ii} & & 1 \end{bmatrix}$$

and in this case $a_{ii} = 1$ throughout. Thus

$$\mathbf{x} = \mathbf{M}_{n-1}\ \mathbf{M}_{n-2} \ldots \mathbf{M}_1\,\mathbf{c}\ .$$

The solution vector $\mathbf{x}$ can be calculated in parallel mode by use of the recursive doubling procedure and thus $0(\log_2 n)$ time steps are carried out in accordance with the Theorem.

To be more specific we now show the details for the $R\langle 4,2\rangle$ system. The recurrence relations are

$$\begin{aligned} x_1 &= c_1 \\ x_2 &= c_2 + a_{21}\,x_1 \\ x_3 &= c_3 + a_{31}\,x_1 + a_{32}\,x_2 \\ x_4 &= c_4 \qquad\qquad + a_{42}\,x_2 + a_{43}\,x_3, \end{aligned}$$

or

$$\mathbf{x} = \mathbf{c} + \mathbf{A}\,\mathbf{x}\ ,$$

where

$$\mathbf{A} := \begin{bmatrix} 0 & 0 & 0 & 0 \\ a_{21} & 0 & 0 & 0 \\ a_{31} & a_{32} & 0 & 0 \\ 0 & a_{42} & a_{43} & 0 \end{bmatrix} .$$

Thus

$$\mathbf{L} := \begin{bmatrix} 1 & 0 & 0 & 0 \\ -a_{21} & 1 & 0 & 0 \\ -a_{31} & -a_{32} & 1 & 0 \\ 0 & -a_{42} & -a_{43} & 1 \end{bmatrix} .$$

Hence

$$\mathbf{x} = \mathbf{L}^{-1}\,\mathbf{c} = \mathbf{M}_3\,\mathbf{M}_2\,\mathbf{M}_1\,\mathbf{c} =$$

$$= \begin{bmatrix} 1 & 0 & 0 & 0 \\ 0 & 1 & 0 & 0 \\ 0 & 0 & 1 & 0 \\ 0 & 0 & a_{43} & 1 \end{bmatrix} \begin{bmatrix} 1 & 0 & 0 & 0 \\ 0 & 1 & 0 & 0 \\ 0 & a_{32} & 1 & 0 \\ 0 & a_{42} & 0 & 1 \end{bmatrix} \begin{bmatrix} 1 & 0 & 0 & 0 \\ a_{21} & 1 & 0 & 0 \\ a_{31} & 0 & 1 & 0 \\ 0 & 0 & 0 & 1 \end{bmatrix} \begin{bmatrix} c_1 \\ c_2 \\ c_3 \\ c_4 \end{bmatrix}$$

$$= \begin{bmatrix} 1 & 0 & 0 & 0 \\ 0 & 1 & 0 & 0 \\ 0 & a_{32} & 1 & 0 \\ 0 & (a_{43}a_{32}+a_{42}) & a_{43} & 1 \end{bmatrix} \begin{bmatrix} c_1 \\ a_{21}c_1 + c_2 \\ a_{31}c_1 + c_3 \\ c_4 \end{bmatrix} ,$$

giving

$$\mathbf{x} = \begin{bmatrix} c_1 \\ a_{21}c_1 + c_2 \\ a_{32}(a_{21}c_1 + c_2) + a_{31}c_1 + c_3 \\ (a_{43}a_{32} + a_{42})(a_{21}c_1 + c_2) + a_{43}(a_{31}c_1 + c_3) + c_4 \end{bmatrix} .$$

Parallel evaluation with $m=2, n=4$ gives

$$T_p \leqslant 3\log_2 4 - 1 = 5$$

time unit steps. In fact $T_3 = 5$ steps are necessary with $p=3$ processors. The estimate of the Theorem requires

$$p \leqslant 3{\cdot}4 - 8 = 4 .$$

For further applications see Kuck's work.

3.3.3 General recurrence relations

A general recurrence system $R\langle n,m\rangle$ of the mth order is defined by

$$R\langle n,m\rangle : x_k := H\,[\mathbf{a}_k; x_{k-1}, x_{k-2}, \ldots, x_{k-m}], \quad (1 \leqslant k \leqslant n) \quad ,$$

with m initial conditions $\mathbf{x}_{m+1}, \mathbf{x}_{m+2}, \ldots, x_0$. H is called the recurrence, or recursion, function. $\mathbf{a}_k$ is a vector of parameters which are independent of the x_i. This definition is uniquely suited to SIMD machines. The instructions needed to calculate H are all executed by each processor P.

A simple example of a recurrence system of the first order is given by (3.29) namely

$$\begin{aligned} x_1 &:= c_1 \qquad \text{(initial condition)} \\ x_k &:= a_k x_{k-1} + c_k\,, \qquad 2 \leqslant k \leqslant n \\ &= H\,[\mathbf{a}_k\,;\, x_{k-1}]\ . \end{aligned} \tag{3.40}$$

The parameter vector $\mathbf{a}_k$ is of length 2 and is given by $\mathbf{a}_k = (a_k, c_k)$. The function H is defined by an addition and a multiplication. (3.40) is often written in the form

$$x_k := a_k(1)\,x_{k-1} + a_k(2)\ . \tag{3.41}$$

The length of $\mathbf{a}_k$ is problem dependent, but for a given problem constant for each k.

Example

(3.42) is a nonlinear recurrence of the first order.

$$x_k := H[\mathbf{a}_k; x_{k-1}] = \frac{a_k(1) + a_k(2)\,x_{k-1}}{a_k(3) + a_k(4)\,x_{k-1}}\ , \tag{3.42}$$

$a_k(j) \in \mathbb{R}$ and in this case $\mathbf{a}_k \in \mathbb{R}^4$.

Many recurrence relations are not associative. (3.40) is an example Such relations cannot be resolved directly by the log-sum algorithm. However there often exist *companion-functions* G associated with H, which possess associative properties. If such functions can be found the recursive doubling procedure can be applied and the parallelisation of the recurrence is then effectively reduced to the construction of a companion function.

Let us reconsider in this connection the set (3.40). We note that

$$\begin{aligned} x_{k+1} &= H(\mathbf{a}_{k+1}; x_k) = a_{k+1}\,x_k + c_{k+1} \\ x_{k+2} &= H(\mathbf{a}_{k+2}; x_{k+1}) = H(\mathbf{a}_{k+2}; H(\mathbf{a}_{k+1}; x_k)) \\ &= (a_{k+2}a_{k+1})x_k + (a_{k+2}\,c_{k+1} + c_{k+2}). \end{aligned} \tag{3.43}$$

The last may be written

$$x_{k+2} = H(\mathbf{a}(k+2, k+1); x_k)$$

with

$$\mathbf{a}(k+2, k+1) = (a_{k+2}a_{k+1}, a_{k+2}c_{k+1} + c_{k+2}).$$

That is to say, we have introduced a new parameter vector $\mathbf{a}(k+2, k+1)$ which is independent of each x_j. x_{k+1} and x_{k+2} can thus be calculated in parallel with a SIMD machine.

Definition (3.44)
A function G is said to be a *companion function* to the recurrence function H when, for all $x \in \mathbb{R}$ abd for all $\mathbf{a}, \mathbf{b} \in \mathbb{R}^p$ we have

$$H(\mathbf{a}; H(\mathbf{b}, x)) = H(G(\mathbf{a}, \mathbf{b}); x) \qquad (3.45)$$

where $G : \mathbb{R}^p \times \mathbb{R}^p \to \mathbb{R}^p$

The following Theorem holds.

Theorem (3.46)
Every companion function is associative with respect to its recurrence function H. That is to say, for all $x \in \mathbb{R}$ and $\mathbf{a}, \mathbf{b}, \mathbf{c} \in \mathbb{R}^p$ we have

$$H(G(\mathbf{a}, G(\mathbf{b}, \mathbf{c})); x) = H(G(G(\mathbf{a},\mathbf{b}), \mathbf{c}); x) \quad . \qquad (3.47)$$

Proof
From the definition of G we have

$$\begin{aligned} H(\mathbf{a}; H(\mathbf{b}; H(\mathbf{c}; x))) &= H(\mathbf{a}; H(G(\mathbf{b}, \mathbf{c}); x)) \\ &= H(G(\mathbf{a}, G(\mathbf{b}, \mathbf{c})); x). \end{aligned}$$

Also

$$\begin{aligned} H(\mathbf{a}; H(\mathbf{b}; H(\mathbf{c}; x))) &= H(G(\mathbf{a}, \mathbf{b}); H(\mathbf{c}; x)) \\ &= H(G(G(\mathbf{a}, \mathbf{b}), \mathbf{c}); x), \end{aligned}$$

and the Theorem is proved.

A further important property is given by the following

Theorem (3.48)
For every recurrence problem of form

$$x_k := H(\mathbf{a}_k; x_{k-1}), \qquad 1 \leqslant k \leqslant n \quad ,$$

with given x_0,

$$x_k = H(\mathrm{a}(k, j+1); x_j) \qquad 0 \leqslant j < k \leqslant n, \tag{3.49}$$

where

$$\mathrm{a}(k,r) = \begin{cases} \mathrm{a}_k \text{ for } r=1, \\ G(\mathrm{a}_k, G(\mathrm{a}_{k-1}, \ldots, G(\mathrm{a}_{k-r+2}, \mathrm{a}_{k-r+1}) \ldots) \text{ for } r > 1 \ . \end{cases}$$

The proof follows directly from the definition of G. The existence of G permits the construction of a parallel algorithm using the log-sum algorithm for the original problem.

Example
According to the Theorem (3.48) we have

$$x_4 = H(\mathrm{a}_4; H(\mathrm{a}_3; H(\mathrm{a}_2; H(\mathrm{a}_1; x_0)))) =$$
$$= H(G(\mathrm{a}_4, G(\mathrm{a}_3, G(\mathrm{a}_2, \mathrm{a}_1))); x_0)$$

The associativity of G gives finally:

$$x_4 = H(G(G(\mathrm{a}_4, \mathrm{a}_3), G(\mathrm{a}_2, \mathrm{a}_1)); x_0).$$

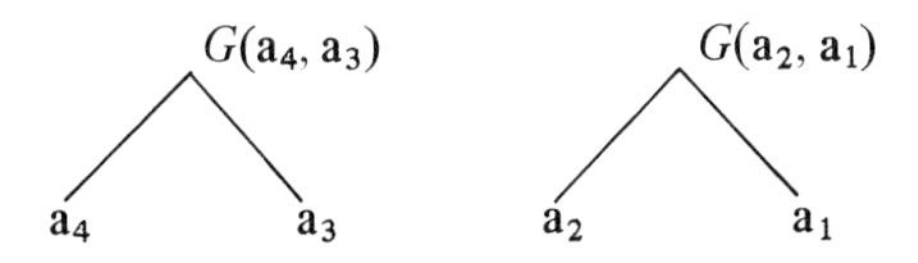

The expressions $G(\mathrm{a}_4, \mathrm{a}_3)$, $G(\mathrm{a}_2, \mathrm{a}_1)$ are computed in parallel. This preliminary step can be applied for each x_k, $1 \leqslant k \leqslant n$ and we obtain the so-called *First Order Recursion Algorithm* (FORA) for the parallel computation of the sequence $x_1, x_2, \ldots, x_n$. With n processors $\log_2 n$ computations are needed for the G functions and one further computation for H.

From Theorem (3.48) it follows in the case of first order recurrence that

$$x_k := H(\mathrm{a}(k, 1); x_0), \qquad 1 \leqslant k \leqslant n \ . \tag{3.50}$$

The FORA algorithm can be applied *directly* to solve such a set, explicitly

$$x_k := a_k x_{k-1} + c_k, \ 1 \leqslant k \leqslant n \ , \tag{3.51}$$

with given x_0. For (3.51) is of the form

$$H(\mathrm{a}_k; x_{k-1}) = \mathrm{a}_k(1)\, x_{k-1} + \mathrm{a}_k(2) \ .$$

Now

$$H(\mathrm{a};H(\mathrm{b};x)) = \mathrm{a}(1)\,(\mathrm{b}(1)x+\mathrm{b}(2)) + \mathrm{a}(2)$$

$$= H((\mathrm{a}(1)\,\mathrm{b}(1), \mathrm{a}(1)\,\mathrm{b}(2) + \mathrm{a}\,(2));x) \quad ,$$

and so

$$G(\mathrm{a}, \mathrm{b}) = (\mathrm{a}(1)\,\mathrm{b}(1), \mathrm{a}(1)\,\mathrm{b}(2) + \mathrm{a}(2)).$$

It must be noted that the parallelisation of nonlinear recurrences does not always yield a speed advantage. Consequently each case must be considered on its individual merits.

A recurrence system $R\,\langle n, m\rangle$ of order m with n equations can be calculated by parallel processing in $0(\log_2 n)$ steps if the equations are linear (see Kuck [12]). Such an equation has the relational form

$$R\,\langle n, m\rangle: x_k = H(x_{k-1}, x_{k-2}, \ldots, x_{k-m}), \quad 1 \leqslant k \leqslant n \quad .$$

Kung [13] has shown that for rational functions H of degree $s > 1$, the use of algebraic algorithms to achieve parallel computation cannot increase the computation speed of $R\,\langle n, m\rangle$ by more than a constant factor.

An example of a first order nonlinear recurrence relation is provided by the well-known Newton algorithm for the calculation of the square root of a real positive number a. This is

$$x_{n+1} = \tfrac{1}{2}\left(x_n + \frac{a}{x_n}\right) = \frac{x_n^2 + a}{2x_n} \,,$$

and the procedure is to start with an initial approximation x_0 to $a^{1/2}$ which is greater than $a^{1/2}$, and to calculate successive approximations x_n until agreement to within the required accuracy is obtained.

The recurrence can indeed be transformed into a linear one by putting

$$x_n = a^{1/2} \coth y_n ,$$

which produces the *linear* recurrence

$$y_{n+1} = 2y_n \quad .$$

Parallel calculation of a set of values y_k $(k=1, 2, \ldots, n)$ is obviously possible but the practical limitation is that to use the transformation means knowing in advance the value $a^{1/2}$ sought!

Another nonlinear example is the procedure of the arithmetic–geometric mean which can be used with suitable starting values to produce successive approximations to complete elliptic integrals. The basic recurrence which gives the procedure its name is usually written as

$$x_{n+1} = \tfrac{1}{2}(x_n + y_n); \quad y_{n+1} = (x_n y_n)^{\frac{1}{2}}$$

and calculation continues until x_n and y_n agree to prescribed accuracy. The formal similarity to the Newton algorithm can be seen by putting $z_n = x_n/y_n$, which enables the procedure to be written

$$z_{n+1} = \tfrac{1}{2}(z_n^{\frac{1}{2}} + z_n^{-\frac{1}{2}}) \quad .$$

If there is a simple transformation that converts this to a linear recurrence it is not obvious.

The procedure is fully described in [14].

4

Development of special algorithms

For the purpose of this chapter the parallel machine will be assumed, unless otherwise stated, to satisfy the following restrictions.

(a) All processors are available at every moment, but limits on the number of processors are given.
(b) Each processor can carry out each of the four basic arithmetic operations in a single time unit. Moreover, in a prescribed time step all processors execute the same instruction. Thus we are assuming an SIMD machine.
(c) There are no limitations on data or storage.

(c) is somewhat superfluous for the discussion of the algorithms since in practical cases one has anyway to take into account the fact that the number of processors available *is* limited and special data paths must be considered.

4.1 Solution of Systems of linear Equations

General remarks

Chen and Kuck [15], Heller [8], and Borodin and Munro [16] have shown that if $p=0(n^3)$ processors are available a triangular system of n equations of form $\mathbf{L}\,\mathbf{x} = \mathbf{b}$ can be resolved in $0(\log_2^2 n)$ time steps. In the following section algorithms will be presented for the solution of $\mathbf{L}\,\mathbf{x} = \mathbf{b}$ with a dense triangular matrix $\mathbf{L}$.

4.1.1 Dense triangular systems

Two algorithms due to Sameh and Brent [17] for the solution of a dense triangular system of equations will be discussed.

Algorithm 1

It is supposed that the system to be solved is of the form $\mathbf{L}\,\mathbf{x} = \mathbf{b}$ with $\mathbf{L}$ a lower triangular matrix $[l_{ij}\colon i,j=1, 2, \ldots, n]$. The inverse is given by

$$\mathbf{L}^{-1} = \mathbf{M}_n \mathbf{M}_{n-1} \ldots \mathbf{M}_1,$$

where

$$\mathbf{M}_i = \begin{bmatrix} 1 & & & & & & \\ & \ddots & & & & 0 & \\ & & 1 & & & & \\ & & & 1/l_{ii} & & & \\ & & & -l_{i+1,i}/l_{ii} & 1 & & \\ & 0 & & \vdots & & \ddots & \\ & & & -l_{n,i}/l_{ii} & & & 1 \end{bmatrix} \tag{4.1}$$

$(i=1, 2, \ldots, n)$.

The reader has encountered this already in previous chapters. It is due to Householder [18] but the proof can be carried out by elementary means. The solution of the equation is given by

$$\mathbf{x} = \mathbf{M}_n \mathbf{M}_{n-1} \ldots \mathbf{M}_1 \mathbf{b} \ . \tag{4.2}$$

(4.2) can be realised in $\mu = \log_2 n$ steps in each of which a matrix multiplication must be carried out. The final multiplication by $\mathbf{M}_n$ requires another scalar multiplication. Since the matrices are sparse the algorithm does not require a full matrix multiplication at any stage.

To develop the algorithm we now define

$$\mathbf{M}_i^{(0)} := \mathbf{M}_i \qquad (i = 1, 2, \ldots, n-1) \tag{4.3}$$

$$\mathbf{b}^{(0)} := \mathbf{b}$$

$$s_j := 2^j \qquad (j = 0, 1, \ldots \mu = \log_2 n)$$

$$\mathbf{M}_i^{(j+1)} := \mathbf{M}_{2i+1}^{(j)} \mathbf{M}_{2i}^{(j)} \qquad \begin{cases} j = 0, 1, \ldots, \mu-2 \\ i = 1, 2, \ldots, \dfrac{n}{s_{j+1}} - 1, \end{cases}$$

$$\mathbf{b}^{(j+1)} := \mathbf{M}_1^{(j)} \mathbf{b}^{(j)} \qquad (j=0, 1, \ldots, \mu-1) \ .$$

It will now be shown that each matrix $\mathbf{M}_i^{(j)}$ has the form

$$\mathbf{M}_i^{(j)} = \left[\begin{array}{c|c|c} \mathbf{I}_i^{(j)} & \mathbf{0} & \mathbf{0} \\ \hline \mathbf{0} & \hat{\mathbf{L}}_i^{(j)} & \mathbf{0} \\ \hline \mathbf{0} & \mathbf{R}_i^{(j)} & \hat{\mathbf{I}}_i^{(j)} \end{array}\right] \tag{4.4}$$

where $\hat{\mathbf{L}}_i^{(j)}$ is a lower triangular matrix of dimension s_j, $\mathbf{I}_i^{(j)}$ and $\hat{\mathbf{I}}_i^{(j)}$ are unit matrices of dimensions

$$q_i^{(j)} = is_j - 1 \quad ,$$

and

$$r_i^{(j)} = n+1-(i+1)s_j \tag{4.5}$$

respectively.

The proof is carried out easily by induction. For $j=0$, $s_0=1$, $q_i^{(0)}=i-1$, $r_i^{(0)} = n-i$. To Make the passage from j to $j+1$ let the matrix element $\mathbf{R}_i^{(j)}$ in $\mathbf{M}_i^{(j)}$ be partitioned as follows

$$\mathbf{R}_i^{(j)} = \begin{bmatrix} \mathbf{U}_i^{(j)} \\ \mathbf{V}_i^{(j)} \end{bmatrix} \quad , \quad (i=1, \ldots, \frac{n}{s_j} - 2) \quad , \tag{4.6}$$

where $\mathbf{U}_i^{(j)}$ is an $s_j \times s_j$ matrix. Then, by (4.3) ,

$$\mathbf{M}_i^{(j+1)} = \begin{bmatrix} \mathbf{I}_{2i}^{(j)} & & & \\ & \mathbf{I}_{s_j} & & \\ & & \hat{\mathbf{L}}_{2i+1}^{(j)} & \\ & & \mathbf{R}_{2i+1}^{(j)} & \hat{\mathbf{I}}_{2i+1}^{(j)} \end{bmatrix} \begin{bmatrix} \mathbf{I}_{2i}^{(j)} & & & \\ & \hat{\mathbf{L}}_{2i}^{(j)} & & \\ & \mathbf{U}_{2i}^{(j)} & \mathbf{I}_{s_j} & \\ & \mathbf{V}_{2i}^{(j)} & & \hat{\mathbf{I}}_{2i+1}^{(j)} \end{bmatrix} =$$

$$= \begin{bmatrix} \mathbf{I}_{2i}^{(j)} & & & \\ & \hat{\mathbf{L}}_{2i}^{(j)} & & \\ & \hat{\mathbf{L}}_{2i+1}^{(j)}\mathbf{U}_{2i}^{(j)} & \hat{\mathbf{L}}_{2i+1}^{(j)} & \\ & \mathbf{R}_{2i+1}^{(j)}\mathbf{U}_{2i}^{(j)}+\mathbf{V}_{2i}^{(j)} & \mathbf{R}_{2i+1}^{(j)} & \hat{\mathbf{I}}_{2i+1}^{(j)} \end{bmatrix}$$

$$(i=1, 2, \ldots, \frac{n}{s_{j+1}} - 1).$$

Note: Since the matrices are partitioned in a special way, the product can be calculated by normal matrix multiplication.

To continue, $\mathbf{M}_i^{(j+1)}$ is thus of the form

$$\mathbf{M}_i^{(j+1)} = \left[\begin{array}{c|c|c} \mathbf{I}_i^{(j+1)} & & \\ \hline & \hat{\mathbf{L}}_i^{(j+1)} & \\ \hline & \mathbf{R}_i^{(j+1)} & \hat{\mathbf{I}}_i^{(j+1)} \end{array}\right] ,$$

with

$$\mathbf{I}_i^{(j+1)} = \mathbf{I}_{2i}^{(j)} .$$

Thus

$$q_i^{(j+1)} = q_{2i}^{(j)} = 2i\, s_j - 1 = i\, s_{j+1} - 1 ,$$

$$\hat{\mathbf{I}}_i^{(j+1)} = \hat{\mathbf{I}}_{2i+1}^{(j)} ,$$

and

$$r_i^{(j+1)} = r_{2i+1}^{(j)} = n+1-(2i+2)\, s_j = n+1-(i+1)\, s_{j+1} ,$$

$$\mathbf{R}_i^{(j+1)} = \left[\begin{array}{c|c} \mathbf{R}_{2i+1}^{(j)} \mathbf{U}_{2i}^{(j)} + \mathbf{V}_{2i}^{(j)} & \mathbf{R}_{2i+1}^{(j)} \end{array}\right] \tag{4.7}$$

$$(j=0, 1, \ldots, \mu-2; i=1, 2, \ldots, \frac{n}{s_{j+1}}-1) ,$$

and

$$\mathbf{L}_i^{(j+1)} = \left[\begin{array}{c|c} \hat{\mathbf{L}}_{2i}^{(j)} & \\ \hline \hat{\mathbf{L}}_{2i+1}^{(j)} \mathbf{U}_{2i}^{(j)} & \hat{\mathbf{L}}_{2i+1}^{(j)} \end{array}\right] \tag{4.8}$$

$$(j=0, 1, \ldots, \mu-2;\ i=1, 2, \ldots \frac{n}{s_{j+1}}-1) .$$

Thus $\hat{\mathbf{L}}_i^{(j+1)}$ is again a triangular matrix and the induction is complete Q.E.D.

In particular $\mathbf{M}_1^{(j)}$ has the form

$$\mathbf{M}_1^{(j)} = \left[\begin{array}{c|c|c} \mathbf{I}_{s_j-1} & & \\ \hline & \hat{\mathbf{L}}_1^{(j)} & \\ \hline & \mathbf{R}_1^{(j)} & \mathbf{I}_{n+1-2s_j} \end{array}\right] , \qquad (j=0, 1, \ldots, \mu-2).$$

If $\mathbf{b}^{(j)}$ be correspondingly partitioned into partial vectors of length s_j-1, s_j and $n+1-2s_j$ we have

$$\mathbf{b}^{(j)} = \begin{bmatrix} \mathbf{g}_1^{(j)} \\ \mathbf{g}_2^{(j)} \\ \mathbf{g}_3^{(j)} \end{bmatrix} \qquad (j=0, 1, \ldots, \mu-1) \tag{4.9}$$

and hence

$$\mathbf{b}^{(j+1)} = \mathbf{M}_1^{(j)}\,\mathbf{b}^{(j)} = \begin{bmatrix} \mathbf{g}_1^{(j)} & & \\ \hat{\mathbf{L}}_1^{(j)} & \mathbf{g}_2^{(j)} & \\ \mathbf{R}_1^{(j)} & \mathbf{g}_2^{(j)} & + \ \mathbf{g}_3^{(j)} \end{bmatrix} \tag{4.10}$$

$$(j=0, 1, \ldots, \mu-1) \ .$$

For the solution of $\mathbf{L}\,\mathbf{x} = \mathbf{b}$ where $\mathbf{L}$ is the lower triangular matrix

$$\mathbf{L} = [\, l_{ik} \,]_{n \times n}$$

we obtain the following algorithm:

1. (a) Set $\hat{\mathbf{L}}_i^{(0)} := [l_{ii}^{-1}]$; and

$$\mathbf{R}_i^{(0)} := \begin{bmatrix} -l_{i+1,i}/l_{ii} \\ -l_{i+2,i}/l_{ii} \\ \vdots \\ -l_{n,i}/l_{ii} \end{bmatrix} \qquad (i=1, 2, \ldots, n-1)$$

   (b) Set $\mathbf{b}^{(0)} := \mathbf{b}$.

2. Repeat the following sequence for $j=0, 1, 2, \ldots, \mu-2$ ($\mu := \log_2 n$):
   (a) Put $s_j := 2^j$
   (b) Partition $\mathbf{R}_i^{(j)}$ as in (4.6) and $\mathbf{b}^{(j)}$ as in (4.9).
   (c) Calculate $\hat{\mathbf{L}}_i^{(j+1)}$ as in (4.8) and $\mathbf{R}_i^{(j+1)}$ as in (4.7) ($i=1, 2, \ldots, n/s_{j+1}-1$), and $\mathbf{b}^{(j+1)}$ as in (4.10).

3. Partition $\mathbf{b}^{(\mu-1)}$ as in (4.9) ($s_{\mu-1} := 2^{\mu-1}$) and calculate $\mathbf{b}^{(\mu)}$ by (4.10).

4. The solution $\mathbf{x}$ is given by

$$\mathbf{x} = (b_1^{(\mu)}, b_2^{(\mu)}, \ldots, b_{n-2}^{(\mu)}, b_{n-1}^{(\mu)}, l_{nn}^{-1} b_n^{(\mu)})^{\mathrm{T}}$$

where $\mathbf{b} = (b_1, b_2, \ldots, b_n)^{\mathrm{T}}$.

Step 4 effectively carries out multiplication by the matrix $\mathbf{M}_n$.

Under the assumption we have made about the computer each matrix $\mathbf{M}_i$ can be constructed in two steps, division and subtraction, if $n-i+1$ processors are used, this being the number of non-null elements in the ith column. Thus the construction of all the matrices $\mathbf{M}_i$ $(i=1, 2, \ldots, n)$ can be carried out in two steps when $\sum_{i=1}^{n} (n-i+1) = \frac{1}{2}n(n+1)$ processors are available. The following Theorem is due to Sameh and Brent [17].

Theorem

The triangular system of equations $\mathbf{L}\,\mathbf{x} = \mathbf{b}$, where $\mathbf{L}$ is a lower triangular matrix of order n, can be solved in

$$T = \tfrac{1}{2} \log_2^2 n + \tfrac{3}{2} \log_2 n + 3$$

time steps and needs at most

$$p = \frac{15}{1204} n^3 + 0(n^2)$$

processors.

Observation

$\mathbf{L}^{-1}$ can be calculated in the same time T using at most

$$p = (21n^3 + 60n^2)/128$$

processors.

Algorithm 2

In [17] Sameh and Brent give another algorithm for the solution of $\mathbf{A}\,\mathbf{x} = \mathbf{b}$ in which $\mathbf{A}$ is a complete lower triangular unit matrix. Effectively the solution of a system $\mathbf{L}\,\mathbf{x} = \mathbf{b}$ is carried out by

$$\mathbf{x} = \mathbf{L}^{-1}\,\mathbf{b} = \mathbf{D}_{\mu-1} \ldots \mathbf{D}_0\,\mathbf{b}$$

which entails $\mu = \log_2 n$ special matrices $\mathbf{D}_i$. Starting with $\mathbf{L}^{(0)} := \mathbf{L}$ we can so choose the $\mathbf{D}_i$ that the intermediate matrices

$$\mathbf{L}^{(j)} := \mathbf{D}_{j-1} \ldots \mathbf{D}_0 \, \mathbf{L}^{(0)} \qquad (j=1, 2, \ldots, \mu) \tag{4.11}$$

have the form

$$\mathbf{L}^{(j)} = \begin{bmatrix} \mathbf{I}_{s_j} & & & & \\ \mathbf{H}_{21}^{(j)} & \mathbf{I}_{s_j} & & & \\ \mathbf{H}_{31}^{(j)} & \mathbf{H}_{32}^{(j)} & \mathbf{I}_{s_j} & & \\ \vdots & \vdots & & \ddots & \\ \mathbf{H}_{\frac{n}{s_j},1}^{(j)} & \mathbf{H}_{\frac{n}{s_j},2}^{(j)} & \ldots & \mathbf{H}_{\frac{n}{s_j},\frac{n}{s_j}-1}^{(j)} & \mathbf{I}_{s_j} \end{bmatrix} . \tag{4.12}$$

First we note that $\mathbf{L}^{(0)}$ is already of the form (4.12). Next we show that if (4.12) holds for $\mathbf{L}^{(j)}$ $(j=0, 1, \ldots, \mu-1)$ it holds also for $\mathbf{L}^{(j+1)} = \mathbf{M}_j \, \mathbf{L}^{(j)}$. The matrices $\mathbf{M}_j$ are easily obtained by using the simple facts that (i) the inverse of a matrix of form $\mathbf{A} = \begin{bmatrix} 1 & 0 \\ a & 1 \end{bmatrix}$ is given by $\mathbf{A}^{-1} = \begin{bmatrix} 1 & 0 \\ -a & 1 \end{bmatrix}$, as can easily be verified by multiplication; and (ii) the diagonal elements of an $n \times n$ triangular matrix $\mathbf{A} = [a_{ik}]$ can be normalised to unity by premultiplication by the diagonal matrix $\mathbf{D} := \operatorname{diag}\,[a_{11}^{-1}, a_{22}^{-1}, \ldots, a_{nn}^{-1}]$. (i) and (ii) naturally hold also when the elements of the matrices are themselves matrices. Hence

$$\mathbf{D}_j = \operatorname{diag}\,(\mathbf{L}_1^{(j)}, \mathbf{L}_2^{(j)}, \ldots, \mathbf{L}_{n/s_j}^{(j)}), \tag{4.13}$$

with

$$\mathbf{L}_i^{(j)} = \left[\begin{array}{c|c} \mathbf{I}_{s_j} & \mathbf{0} \\ \hline -\mathbf{H}_{2i,\,2i-1}^{(j)} & \mathbf{I}_{s_j} \end{array}\right] \qquad \begin{array}{l} (j=0, 1, \ldots, n-1; \\ i=0, 1, \ldots, n/s_j) \end{array} .$$

For the implementation of the algorithm it is necessary to determine the relationship satisfied by the matrices $\mathbf{H}_{ik}^{(j)}$.

Using the abbreviation $\sigma := n/s_j$ we have:

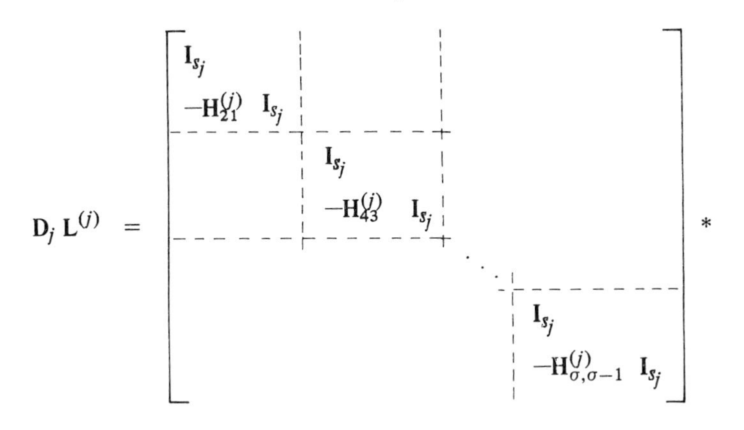

$$
\mathbf{D}_j \mathbf{L}^{(j)} = \left[\text{as above}\right] *
\begin{bmatrix}
\mathbf{I}_{s_j} & & & & & & \\
\mathbf{H}_{21}^{(j)} & \mathbf{I}_{s_j} & & & & & \\
\mathbf{H}_{31}^{(j)} & \mathbf{H}_{32}^{(j)} & \mathbf{I}_{s_j} & & & & \\
\mathbf{H}_{41}^{(j)} & \mathbf{H}_{42}^{(j)} & \mathbf{H}_{43}^{(j)} & \mathbf{I}_{s_j} & & & \\
\vdots & \vdots & & & \ddots & & \\
\mathbf{H}_{\sigma-1,1}^{(j)} & \mathbf{H}_{\sigma-1,2}^{(j)} & & & & \mathbf{I}_{s_j} & \\
\mathbf{H}_{\sigma,1}^{(j)} & \mathbf{H}_{\sigma,2}^{(j)} & & & & \mathbf{H}_{\sigma,\sigma-1}^{(j)} & \mathbf{I}_{s_j}
\end{bmatrix}
$$

$$
= \begin{bmatrix}
\mathbf{I}_{s_{j+1}} & & & \\
\mathbf{H}_{21}^{(j+1)} & \mathbf{I}_{s_{j+1}} & & \\
\vdots & & \ddots & \\
\mathbf{H}_{\frac{\sigma}{s_{j+1}},1}^{(j+1)} & \cdots & & \mathbf{I}_{s_{j+1}}
\end{bmatrix} = \mathbf{L}^{(j+1)}
$$

$(j = 0, 1, \ldots, \mu-1)$

as required. Thus

$$\mathbf{H}_{ik}^{(j+1)} = \left[\begin{array}{c|c} \mathbf{I}_{s_j} & \mathbf{0} \\ \hline -\mathbf{H}_{2i,2i-1}^{(j)} & \mathbf{I}_{s_j} \end{array}\right] * \left[\begin{array}{c|c} \mathbf{H}_{2i-1,2k-1}^{(j)} & \mathbf{H}_{2i-1,2k}^{(j)} \\ \hline \mathbf{H}_{2i,2k-1}^{(j)} & \mathbf{H}_{2i,2k}^{(j)} \end{array}\right] \tag{4.14}$$

for $j=0, 1, \ldots, \mu-1; k=1, 2, \ldots, (n/s_{j+1})-1$; and $i=k+1, k+2, \ldots, n/s_{j+1}$.

For the intermediate solutions

$$\mathbf{b}^{(j)} = \mathbf{D}_{j-1}\,\mathbf{D}_{j-2}\ldots\mathbf{D}_0\,\mathbf{b} = \mathbf{D}_{j-1}\,\mathbf{b}^{(j-1)}$$

we get by corresponding partition of $\mathbf{b}^{(j)}$ into n/s_j vectors $b_i^{(j)}$ of length s_j

$$\mathbf{b}_i^{(j+1)} = \begin{bmatrix} \mathbf{b}_{2i-1}^{(j)} \\ \mathbf{b}_{2i}^{(j)} - \mathbf{H}_{2i,2i-1}^{(j)}\;\mathbf{b}_{2i-1}^{(j)} \end{bmatrix}, \qquad \begin{array}{l}(j=0,\ldots,\mu-1;\\ i=1,\ldots,n/s_{j+1}),\end{array} \tag{4.15}$$

where $\mathbf{b}_1^{(\mu)} = \mathbf{x}$.

Algorithm 2 is in form very similar to Algorithm 1. Nevertheless it is fundamentally different and can not be transformed into Algorithm 1 by formal operations.

The number of time steps T_p is given in [17] as

$$T_p = \tfrac{1}{2}\log_2^2 n + \tfrac{3}{2}\log_2 n + 1,$$

which is virtually the same as Algorithm 1, for the multiplication of matrices with the same dimensions occupy $\log_2 n$ steps. Algorithm 2, however, requires more processors, namely

$$p = \frac{n^3}{32} + \frac{n^2}{8} \qquad (\text{for } n > 16)$$

since the matrix multiplication requirement is greater. From the practical point of view Algorithm 2, nevertheless, has the advantage of being formally simpler. It can, moreover, be carried over very easily to deal with triangular systems having a band structure, as we shall see later. Since T_p and p are more or less equal for both algorithms the following speed-up and efficiency measures for each are valid:

$$S_p = 0\left(\frac{n^2}{\log_2^2 n}\right), \qquad E_p = 0\left(\frac{1}{n\log_2^2 n}\right).$$

4.1.2 Triangular Systems with Band Structure

The algorithms now considered are due to Chen, Kuck and Sameh [19]. The number of processors is prescribed, that is forms a part of the algorithm, whereas previously the number of processors was determined by the algorithm.

First we consider systems of linear equations having the form

$$\mathbf{L}\,\mathbf{x} = \mathbf{b} \tag{4.16}$$

where $\mathbf{L} = [l_{ik}]$ is a lower triangular matrix of order n with unit diagonal terms $l_{ii} = 1$, $i=1, 2, \ldots, n$. If the bandwidth of $\mathbf{L}$ is m, that is to say $l_{ik} = 0$ for $i-k > m$, then the system is equivalent to a linear recurrence $R\langle n, m\rangle$ of order m. It is here indeed that the greatest number of important problems is to be found: one seeks to solve the recurrences

$$x_1 := \phi_1, \quad x_2 := l_{21}\,x_1 + \phi_2 \;, \quad \ldots$$

$$x_i := \sum_{k=1}^{i-1} l_{ik}\,x_k + \phi_i \;,$$

or

$$\mathbf{x} := \hat{\mathbf{L}}\,\mathbf{x} + \mathbf{b}$$

with

$$\hat{\mathbf{L}} = [\hat{l}_{ik}], i, k=1, 2, \ldots, n;\; \hat{l}_{ii} = 0,$$

$$\mathbf{x} = (x_1, x_2, \ldots, x_n)^{\mathrm{T}}, \quad \mathbf{b} = (\phi_1, \phi_2, \ldots, \phi_n)^{\mathrm{T}}.$$

Finally, putting $\mathbf{L} := \mathbf{I} - \hat{\mathbf{L}}$ we recover equation system (4.16), namely

$$\mathbf{L}\,\mathbf{x} = \mathbf{b}\;.$$

Without loss of generality we suppose that the number of processors p has the form

$$p = km \tag{4.17}$$

where k is an integer such that

$$2 \leqslant k \leqslant \frac{n}{m}$$

and n is a multiple of m. Further it will be supposed that $n=sk$. $\mathbf{L}$ will first be partitioned as follows:

$$\mathbf{L} = \begin{bmatrix} \mathbf{L}_1 & & & & \mathbf{0} \\ \mathbf{R}_1 & \mathbf{L}_2 & & & \\ & \mathbf{R}_2 & \mathbf{L}_3 & & \\ & & \ddots & \ddots & \\ \mathbf{0} & & & \mathbf{R}_{k-1} & \mathbf{L}_k \end{bmatrix} = \tag{4.18}$$

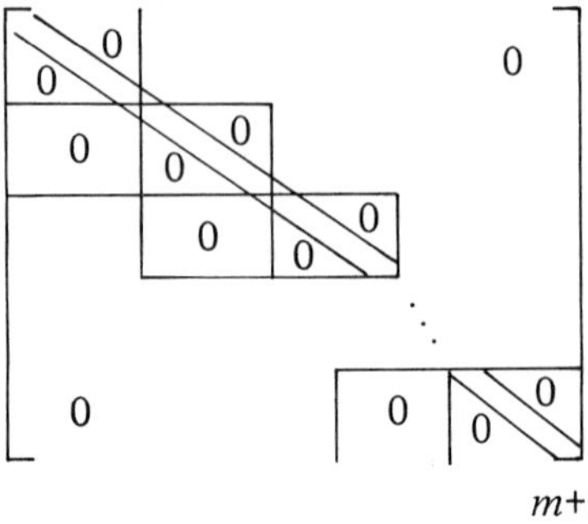

The matrices $\mathbf{L}_i$ (i=1, 2, . . . , k) in (4.18) are lower triangular with unit diagonal, bandwidth m and order $s = n/k$. The matrices $\mathbf{R}_i$ (i=1, 2, . . . , k–1) are upper triangular with non-zero elements r_{jl} only for $l-j \geqslant s-m$, namely

$\mathbf{R}_i =$

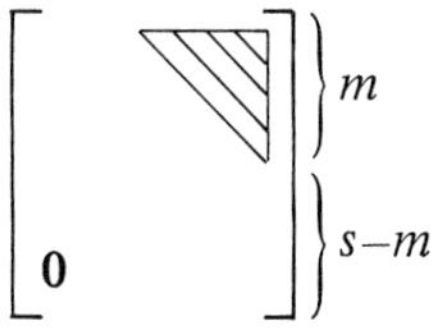

Premultiplying both sides of $\mathbf{L}\,\mathbf{x} = \mathbf{b}$ by

$$\mathbf{D} := \operatorname{diag}\,[\mathbf{L}_1^{-1}, \mathbf{L}_2^{-1}, \ldots, \mathbf{L}_k^{-1}] \tag{4.19}$$

we get

$$\mathbf{D}\,\mathbf{L}\,\mathbf{x} = \mathbf{D}\,\mathbf{b} =: \hat{\mathbf{b}} \tag{4.20}$$

say. If now $\mathbf{x}$, $\mathbf{b}$, $\hat{\mathbf{b}}$ are divided into vectors $\mathbf{x}_i$, $\mathbf{b}_i$, $\hat{\mathbf{b}}_i$ respectively (i=1, 2, . . . , k) each having s elements, then (4.20) can be written in the form

$$\begin{bmatrix} \mathbf{I}_s & & & & \\ \mathbf{G}_1 & \mathbf{I}_s & & & \\ & \mathbf{G}_2 & \mathbf{I}_s & & \\ & & \ddots & \ddots & \\ & & & \mathbf{G}_{k-1} & \mathbf{I}_s \end{bmatrix} \begin{bmatrix} \mathbf{x}_1 \\ \mathbf{x}_2 \\ \mathbf{x}_3 \\ \vdots \\ \mathbf{x}_k \end{bmatrix} = \begin{bmatrix} \hat{\mathbf{b}}_1 \\ \hat{\mathbf{b}}_2 \\ \hat{\mathbf{b}}_3 \\ \vdots \\ \hat{\mathbf{b}}_k \end{bmatrix} \tag{4.21}$$

Now $\hat{\mathbf{b}}_i$ (i=1, 2, . . . , k) and $\mathbf{G}_i$ (i=1, 2, . . . , k–1) are obtained by solving the following linear triangular system

$$\begin{aligned} \mathbf{L}_i\,\mathbf{b}_i &= \mathbf{b}_i\ , \qquad && i=1,\ldots,k; \\ \mathbf{L}_{i+1}\,\mathbf{G}_i &= \mathbf{R}_i\ , && i=1,\ldots,k-1\ . \end{aligned} \tag{4.22}$$

The solution can be subdivided into the following steps:

Step 1: Solution of $\mathbf{L}_i\mathbf{b}_i = \hat{\mathbf{b}}_i$
If m processors are used, each system $\mathbf{L}_i\,\mathbf{b}_i = \hat{\mathbf{b}}_i$ (i=1, 2, . . . , k) can be solved in $2(s-1)$ steps if a generalisation of the sequential algorithm (Sameh, Chen and Kuck [12]) is used. Then

$$T^{(1)} = 2(s-1) = 2\left(\frac{mn}{p}-1\right) \tag{4.23}$$

steps are required with p=mk processors in use to solve the k systems

$$\mathbf{L}_i\,\mathbf{b}_i = \hat{\mathbf{b}}_i \qquad (i=1,2,\ldots,k)\ .$$

Step 2: Solution of $\mathbf{L}_{i+1}\,\mathbf{G}_i = \mathbf{R}_i$
In this step the $\mathbf{G}_i$ (i=1, 2, . . . , k–1) are determined. Since each matrix $\mathbf{R}_i$ is of the form $[\mathbf{0},\hat{\mathbf{R}}_i]$ where $\hat{\mathbf{R}}_i$ is an (s,m) matrix, each $\mathbf{G}_i$ must also have the same form, namely $\mathbf{G}_i = [\mathbf{0},\hat{\mathbf{G}}_i]$ where $\hat{\mathbf{G}}_i$ are to be obtained by solving

$$\mathbf{L}_{i+1}\,\hat{\mathbf{G}}_i = \hat{\mathbf{R}}_i\ .$$

Thus there are altogether $m(k-1) = p-m$ linear systems

$$\mathbf{L}\,\mathbf{g} = \mathbf{h} \tag{4.24}$$

to be solved, where $\mathbf{g}$ and $\mathbf{h}$ are the columns of $\mathbf{G}_i$ and $\mathbf{R}_i$ respectively. If we use a single processor for each system (4.24) then all $p-m$ systems are solved in

$$T^{(2)} = 2sm-m-m^2 = 2m^2\,\frac{n}{p}-m-m^2 \tag{4.25}$$

steps provided that $p-m$ processors are available.

In order to decompose the solution vectors $\mathbf{x}_i$ still further we now partition each matrix $\mathbf{G}_i$ as follows:

$$\mathbf{G}_i = \left[\begin{array}{c|c} \mathbf{0} & \mathbf{M}_i \\ \hline \mathbf{0} & \mathbf{H}_i \end{array}\right]\ , \tag{4.26}$$

where $\mathbf{H}_i$ are to be quadratic matrices of order m and therefore $\mathbf{M}_i$ have to be $(s-m,m)$ matrices.

Partitioning the vectors $\mathbf{x}_i$ and $\hat{\mathbf{b}}_i$ of (4.21) correspondingly into

$$\mathbf{x}_i := \begin{bmatrix} \mathbf{y}_i \\ \mathbf{z}_i \end{bmatrix}, \quad \mathbf{b}_i = \begin{bmatrix} \mathbf{u}_i \\ \mathbf{v}_i \end{bmatrix},$$

where $\mathbf{y}_i$, $\mathbf{u}_i$, have $s-m$ elements and $\mathbf{z}_i$, $\mathbf{v}_i$ have m, we are able to write (4.21) as

$$\begin{bmatrix} \mathbf{P} & & & & \\ \mathbf{G}_1 & \mathbf{P} & & & \\ & \mathbf{G}_2 & \mathbf{P} & & \\ & & \ddots & \ddots & \\ & & & \mathbf{G}_{k-1} & \mathbf{P} \end{bmatrix} \begin{bmatrix} \mathbf{q}_1 \\ \mathbf{q}_2 \\ \vdots \\ \mathbf{q}_k \end{bmatrix} = \begin{bmatrix} \mathbf{r}_1 \\ \mathbf{r}_2 \\ \vdots \\ \mathbf{r}_k \end{bmatrix},$$

where

$$\mathbf{P} = \begin{bmatrix} \mathbf{I}_{s-m} & \mathbf{0} \\ \mathbf{0} & \mathbf{I}_m \end{bmatrix},$$

$$\mathbf{q}_i = \begin{bmatrix} \mathbf{y}_i \\ \mathbf{z}_i \end{bmatrix}, \quad \mathbf{r}_i = \begin{bmatrix} \mathbf{u}_i \\ \mathbf{v}_i \end{bmatrix}, \quad (i=1, 2, \ldots, k) .$$

This decomposition allows (4.21) to be regarded as two systems:

(a) The linear recurrences

$$\mathbf{z}_1 := \mathbf{v}_1$$

$$\mathbf{z}_{i+1} + \mathbf{H}_i \mathbf{z}_i = \mathbf{v}_{i+1} \qquad (i=1, 2, \ldots, k-1); \tag{4.27}$$

(b) The update relations

$$\mathbf{y}_1 := \mathbf{u}_1$$

$$\mathbf{Y}_i := \mathbf{u}_i - \mathbf{M}_{i-1} \mathbf{z}_{i-1}, \qquad (i=2, 3, \ldots, k). \tag{4.28}$$

Step 3: Solution of (4.27)
The solution of (4.27) is equivalent to that of the following linear system of order $p=km$

$$\begin{bmatrix} \mathbf{I}_m & & & & \mathbf{0} \\ \mathbf{H}_1 & \mathbf{I}_m & & & \\ & \mathbf{H}_2 & \mathbf{I}_m & & \\ & & \ddots & \ddots & \\ \mathbf{0} & & & \mathbf{H}_{k-1} & \mathbf{I}_m \end{bmatrix} \begin{bmatrix} \mathbf{z}_1 \\ \mathbf{z}_2 \\ \mathbf{z}_3 \\ \vdots \\ \mathbf{z}_k \end{bmatrix} = \begin{bmatrix} \mathbf{v}_1 \\ \mathbf{v}_2 \\ \mathbf{v}_3 \\ \vdots \\ \mathbf{v}_k \end{bmatrix} , \tag{4.29}$$

or, in brief,

$$\mathbf{L}^{(0)} \mathbf{z} = \mathbf{v}^{(0)} \quad .$$

To solve this system with p processors a variant of Sameh and Brent's [17] Algorithm 2 (see section 4.1.1) is used. We explain it briefly here.

(4.29) is transformed into the set

$$\mathbf{L}^{(j)} \mathbf{z} = \mathbf{v}^{(j)} \qquad (j=1, 2, \ldots, \log_2 k) \tag{4.30}$$

where $\mathbf{L}^{(j)}$ has the form

$$\mathbf{L}^{(j)} = \begin{bmatrix} \mathbf{I}_r & & & & \mathbf{0} \\ \mathbf{H}_1^{(j)} & \mathbf{I}_r & & & \\ & \mathbf{H}_2^{(j)} & \mathbf{I}_r & & \\ & & \ddots & \ddots & \\ \mathbf{0} & & & \mathbf{H}_{(p/r)-1}^{(j)} & \mathbf{I}_r \end{bmatrix} \tag{4.31}$$

for $r=2^j m$, $j=1, 2, \ldots, \log_2 k$.

In particular $\mathbf{L}^{(\log_2 k)} = \mathbf{I}_p$. This is obvious if at the jth step ($j=1, 2, \ldots, \log_2 k$) one premultiplies by

$$\mathbf{D}^{(j)} = \operatorname{diag}[\mathbf{L}_1^{(j)}, \mathbf{L}_2^{(j)}, \ldots, \mathbf{L}_{p/r}^{(j)}]$$

with

$$\mathbf{L}_i^{(j)} = \begin{bmatrix} \mathbf{I}_r & \mathbf{0} \\ -\mathbf{H}_{2i-1}^{(j)} & \mathbf{I}_r \end{bmatrix} .$$

The nature of the matrices $\mathbf{H}_i^{(j)}$ can be made more explicit as follows:

$$\mathbf{D}^{(j)}\,\mathbf{L}^{(j)} = \begin{bmatrix} \mathbf{P}_1 & & & 0 \\ & \mathbf{P}_3 & & \\ & & \ddots & \\ 0 & & & \mathbf{P}_{(p/r)-1} \end{bmatrix} \begin{bmatrix} \mathbf{Q}_1 & & & 0 \\ \mathbf{R}_2 & \mathbf{Q}_3 & & \\ & \ddots & \ddots & \\ 0 & & \mathbf{R}_{(p/r)-2} & \mathbf{Q}_{(p/r)-1} \end{bmatrix}$$

where

$$\mathbf{P}_i = \begin{bmatrix} \mathbf{I}_r & 0 \\ \mathbf{H}_i^{(j)} & \mathbf{I}_r \end{bmatrix}, \quad \mathbf{Q}_i = \begin{bmatrix} \mathbf{I}_r, & 0 \\ \mathbf{H}_i^{(j)}, & \mathbf{I}_r \end{bmatrix}, \quad \mathbf{R}_i = \begin{bmatrix} 0 & \mathbf{H}_i^{(j)} \\ 0 & 0 \end{bmatrix}.$$

The product is

$$\begin{bmatrix} \mathbf{I}_{2r} & & \\ \mathbf{H}_1^{(j+1)} & \mathbf{I}_{2r} & \\ \ddots & \ddots & \\ & \mathbf{H}_{(p/2r)-1}^{(j+1)} & \mathbf{I}_{2r} \end{bmatrix} = \mathbf{L}^{(j+1)} \qquad (j=0, 1, \ldots, (\log_2 k)-1),$$

and hence

$$\mathbf{H}_i^{(j+1)} = \mathbf{P}_{2i+1}\,\mathbf{R}_{2i} = \begin{bmatrix} 0 & \mathbf{H}_{2i}^{(j)} & \\ 0 & -\mathbf{H}_{2i+1}^{(j)} & \mathbf{H}_{2i}^{(j)} \end{bmatrix} \tag{4.32}$$

$$(i=1, 2, \ldots, (p/2r)-1, j=0, 1, \ldots, (\log_2 k)-2).$$

Since the matrices $\mathbf{H}_i^{(0)}$ are $m \times m$ it follows that all matrices $\mathbf{H}_i^{(j)}$ have only their last m columns different from zero. Let $\mathbf{v}_i^{(j)}$ denote the ith partial vector of $\mathbf{v}^{(j)}$, so that

$$\mathbf{v}^{(j)} = (\mathbf{v}_1^{(j)}, \mathbf{v}_2^{(j)}, \ldots, \mathbf{v}_{p/r}^{(j)})^{\mathrm{T}}.$$

Then

$$\mathbf{v}_i^{(j+1)} = \mathbf{D}^{(j)}\,\mathbf{v}_i^{(j)} = \begin{bmatrix} \mathbf{v}_{2i-1}^{(j)} \\ \mathbf{v}_{2i}^{(j)} - \mathbf{H}_{2i-1}^{(j)}\,\mathbf{v}_{2i-1}^{(j)} \end{bmatrix}, \tag{4.33}$$

where $j=0, 1, \ldots, (\log_2 k) - 1; i=1, 2, \ldots, n/r$.

The solution is then, since $\mathbf{L}^{(\log_2 k)} = \mathbf{I}_p$,

$$\mathbf{z} = \mathbf{v}^{(\log_2 k)} \quad .$$

Since the $\mathbf{H}_i^{(j)}$ have $r \times m$ non-zero elements it follows that $\mathbf{H}_{2i+1}^{(j)} \mathbf{H}_{2i}^{(j)}$ produces rm vector products of length m. Since a vector product of this size requires $2m-1$ serial steps, then it follows that to carry out the rm vector products using $2r$ processors requires

$$\frac{rm}{2r}(2m-1) = m\,(m-\tfrac{1}{2})$$

parallel steps. The formation of $-\mathbf{H}_{2i+1}^{(j)} \mathbf{H}_{2i}^{(j)}$ requires an additional $\tfrac{1}{2}m$ parallel steps.

Since in (4.32) altogether $(p/2r) - 1$ matrix multiplications are to be carried out parallel calculation requires $2r\ [(p/2r) - 1]$ processors (less than p). If $2r$ processors are available m steps are necessary for each of the products $\mathbf{H}_{2i-1}^{(j)} \mathbf{v}_{2i-1}^{(j)}$, while for the difference $(\mathbf{v}_{2i}^{(j)} - \mathbf{H}_{2i-1}^{(j)} \mathbf{v}_{2i-1}^{(j)})$ a single step will do. Thus (4.33) can be evaluated in $m+1$ steps with $2r(p/2r) = p$ processors. It follows that the solution of (4.29) requires $T^{(3)}$ steps and at most p processors where

$$T^{(3)} = m^2 \sum_{i=0}^{i_1} 1 + (m+1) \sum_{i=0}^{i_2} 1 \qquad (i_1 = \log_2 (p/4m),\, i_2 = \log_2 (p/2m))$$

$$= (m^2 + m+1) \log_2 (p/m) - m^2 . \tag{4.34}$$

Step 4: Solution of the update relation (4.28)
To achieve a final solution of $\mathbf{L}\mathbf{x} = \mathbf{b}$ we still lack the components of the solution vector $\mathbf{x}$ which were incorporated in the $(s-m)$ vectors $\mathbf{y}_i$ $(i=1, 2, \ldots, k)$. According to (4.28) the following relations hold:

$$\mathbf{y}_1 := \mathbf{u}_1$$

$$\mathbf{y}_i := \mathbf{u}_i - \mathbf{M}_{i-1}\, \mathbf{z}_{i-1} \qquad (i=2, \ldots, k) \quad .$$

The multiplications $\mathbf{M}_{i-1}\, \mathbf{z}_{i-1}$ can be implemented in

$$T^{(4)} = \lceil (k-1)(s-m)(2m-1)/p \rceil^{\dagger} < \frac{2mn}{p} - 2m \tag{4.35}$$

steps with at most p processors. Thus, altogether, $T = \sum_{i=1}^{4} T^{(i)}$ steps are necessary

† $\lceil x \rceil$ is the smallest integer greater than or equal to x. $\lfloor x \rfloor$ is the largest integer that does not exceed x.

to solve $\mathbf{L}\,\mathbf{x} = \mathbf{b}$. A bound on T is given by

$$T < \frac{2m^2n}{p} + \frac{4mn}{p} + 0\left(m^2 \log_2 \frac{p}{m}\right) \quad .$$

Note: If $p = 0(m)$ is very much less than n, then $T = 0(m^2\, n/p)$. If $p{=}n$, it follows that $T = 0\,(m^2 \log_2 n)$. If one is interested in only the last m components of $\mathbf{x}$, then Step 4 is not necessary and the overall time reduces by $T^{(4)} \sim 2mn/p$.

The foregoing discussion can be summarised as follows

Theorem (4.36)
Suppose that p $(2m \leqslant p \leqslant n)$ processors are available. Then any recurrence system $R\,\langle n,m\rangle$ of the form $\mathbf{L}\,\mathbf{x} = \mathbf{b}$ can be solved in

$$T_p = \frac{2m^2n}{p} + \frac{3mn}{p} + 0\left(m^2 \log \frac{p}{2m}\right)$$

time steps, with a minimal speed-up $0\ (p/m)$ where $\mathbf{L}$ is a matrix having the special form

$$\mathbf{L} = \begin{bmatrix} \mathbf{L}_1 & & & & \\ \mathbf{R}_1 & \mathbf{L}_2 & & & \\ & \mathbf{R}_2 & \mathbf{L}_3 & & \\ & & \ddots & \ddots & \\ & & & \mathbf{R}_{k-1} & \mathbf{L}_k \end{bmatrix}$$

and $\mathbf{L}_i$ and $\mathbf{R}_i$ are as in (4.18). With the same number p of processors the final m components of $\mathbf{x}$ can be computed in

$$T_p = \frac{2m^2n}{p} + \frac{2mn}{p} + 0\left(m^2 \log \frac{p}{2m}\right)$$

time steps.

4.1.3 The parallel LR-algorithm and the parallel Gauss algorithm

In this section we shall study the well-known LR decomposition, and the Gauss algorithm for the linear system $\mathbf{A}\,\mathbf{x} = \mathbf{b}$, in the context of parallel computation [13]. In particular, tridiagonal or block-diagonal matrices will be considered, which, as already mentioned, are of particular importance in the numerical solution of partial differential equations. A typical problem of this kind is the solution of an elliptic linear partial differential equation by finite differences.

For this purpose another reference will be made to the Laplace or potential equation. We seek a function $u \in C^2\ (G/\mathbb{R})$ which satisfies the differential equation

$$u_{xx} + u_{yy} = 0 \tag{4.37}$$

in $\bar{G}$,† and the boundary conditions

$$u(x, y) = g(x, y) \tag{4.38}$$
$$\text{for all } (x, y) \in \partial G, \dagger$$

where the condition $g \in C^0\ (\partial G/\mathbb{R})$ is satisfied. The defining region $\bar{G}$ of the function will be taken as the unit quadrant in $\mathbb{R}^2$, but what follows holds also for any axially symmetric region in $\mathbb{R}^2$. A finite difference approach requires placing a net across $\bar{G}$ with lines parallel to the axes at a constant mesh width $h=M^{-1}$ in both x and y directions, and a procedure to approximate the value of u at the nodes of the net.

The procedure is to substitute for the partial derivatives the following difference expressions:

$$h^2 u_{xx}(x, y) = u(x-h, y) - 2u(x, y) + u(x+h, y), \tag{4.39}$$
$$h^2 u_{yy}(x, y) = u(x, y-h) - 2u(x, y) + u(x, y+h).$$

Thus we obtain a discrete difference problem whose solution $v(x, y)$ is an approximation to the solution $u(x, y)$ of the continuous problem. We have

$$4v(x, y) - v(x-h, y) - v(x+h, y) - v(x, y-h) - v(x, y+h) = 0, \tag{4.40}$$

for all $(x, y) \in G^0$,† while $v(x, y) = g(x, y)$ for all $(x, y) \in \partial G$.

The function $v(x, y)$ has to be calculated at each of the interior nodes of the finite difference net. The values on the boundary are supposed known in advance. By introducing a single system of indices we obtain finally a system of linear equations to solve for unknown $\mathbf{v}$, of the form

$$\mathbf{A}\,\mathbf{v} = \mathbf{b}$$

where

† ∂G means the boundary of G, $\bar{G}$ is the closed hull of G and G^0 the inside of G.

$$\mathbf{A} := \begin{bmatrix} \mathbf{X} & -\mathbf{I} & 0 & & 0 \\ -\mathbf{I} & \mathbf{X} & -\mathbf{I} & 0 & 0 \\ 0 & -\mathbf{I} & \mathbf{X} & -\mathbf{I} & 0 \\ & \ddots & \ddots & \ddots & \\ 0 & & -\mathbf{I} & \mathbf{X} & -\mathbf{I} \\ 0 & & 0 & -\mathbf{I} & \mathbf{X} \end{bmatrix} \tag{4.41}$$

with dim $\mathbf{A} = n = k^2$, $k{=}M{-}1$, and $\mathbf{X}$ is the tridiagonal matrix

$$\mathbf{X} := \begin{bmatrix} 4 & -1 & 0 & & & 0 \\ -1 & 4 & -1 & 0 & & 0 \\ 0 & -1 & 4 & -1 & 0 & 0 \\ \vdots & \ddots & \ddots & \ddots & \ddots & \vdots \\ 0 & & 0 & -1 & 4 & -1 \\ 0 & & & 0 & -1 & 4 \end{bmatrix} \tag{4.42}$$

with dim $\mathbf{X} = k$.

Parallel algorithms for the solution of such a system can be implemented on either an SIMD processor or an MIMD processor.

In the following discussion it will be supposed that the size of the linear system is m and that $p{=}m$ processors are available, or, equally, that m is the length of a vector for a vector processor. If $p < m$ then all time estimates must be multiplied by m/p. For the parallel Gauss algorithm a so-called parallel Gauss iteration procedure will be introduced which exhibits favourable numerical properties such, for example, as error norm reduction as the iteration continues. The Gauss iteration resolves a tridiagonal matrix system in a time which is independent of m. In what follows we shall develop first the sequential LR-algorithm and the sequential Gauss algorithm in matrix form. The LR-algorithm decomposes the tridiagonal matrix $\mathbf{A}$ into lower and upper triangular matrices $\mathbf{L}$ and $\mathbf{R}$ respectively so that $\mathbf{A} = \mathbf{L}\,\mathbf{R}$.

The sequential LR-algorithm

It is supposed without loss of generality that the diagonal elements a_{kk} of $\mathbf{A} := [a_{ik}]$ are all unity. $\mathbf{A}$ is now decomposed into

$$\mathbf{A} = \mathbf{A}_L + \mathbf{I} + \mathbf{A}_R = \mathbf{L}\,\mathbf{R} \tag{4.43}$$

where

$$\mathbf{L} = \mathbf{A}_L + \mathbf{D}, \mathbf{R} = \mathbf{I} + \mathbf{J} \tag{4.44}$$

and

D : diagonal matrix

I : unit matrix

$\mathbf{A}_L$: consists of the subdiagonals of **A**, namely

$$\mathbf{A}_L := \begin{bmatrix} 0 & & & & 0 \\ a_{21} & 0 & & & \\ 0 & a_{32} & 0 & & \\ & & \ddots & \ddots & \\ 0 & & & a_{n,n-1} & 0 \end{bmatrix} ;$$

$\mathbf{A}_R$: the superdiagonal terms of **A**, namely

$$\mathbf{A}_R := \begin{bmatrix} 0 & a_{12} & 0 & & & & \\ & 0 & a_{23} & 0 & & 0 & \\ & & 0 & a_{34} & 0 & & \\ & & & 0 & \ddots & & \\ & & & & \ddots & \ddots & \\ & & & & & \ddots & 0 \\ & & & & & & a_{n-1,n} \\ 0 & & & & & & 0 \end{bmatrix} .$$

J has the same structure as $\mathbf{A}_R$.

Then

$$\mathbf{L}\,\mathbf{R} = (\mathbf{A}_L + \mathbf{D})\,(\mathbf{I}+\mathbf{J}) = \mathbf{A}_L + \mathbf{A}_L\,\mathbf{J} + \mathbf{D} + \mathbf{D}\,\mathbf{J} .$$

Now the product $\mathbf{A}_L\ \mathbf{J}$ is a diagonal matrix and hence, since the diagonal of **A** has been supposed to have unit elements,

$$\mathbf{A}_L\,\mathbf{J} + \mathbf{D} = \mathbf{I}$$

and

$$\mathbf{D}\,\mathbf{J} = \mathbf{A}_R \quad . \tag{4.45}$$

Therefore

$$(\mathbf{I} - \mathbf{A}_L\,\mathbf{J})\mathbf{J} = \mathbf{A}_R \quad . \tag{4.46}$$

To solve the system $\mathbf{A}\,\mathbf{u} = \mathbf{v}$ we write

$$(\mathbf{A}_L + \mathbf{D})\,(\,\mathbf{I} + \mathbf{J})\,\mathbf{u} = \mathbf{v}$$

and then define $\mathbf{f}$ by means of

$$(\mathbf{A}_L + \mathbf{D})\mathbf{f} = \mathbf{v} \;.$$

Then

$$(\mathbf{I} + \mathbf{A}_L(\mathbf{I} - \mathbf{J}))\,\mathbf{f} = \mathbf{v} \tag{4.47}$$

and

$$(\mathbf{I}{+}\mathbf{J})\mathbf{u} = \mathbf{f} \quad . \tag{4.48}$$

Now suppose that

$$\mathbf{A} := \begin{bmatrix} 1 & s_1 & 0 & & & & 0 \\ t_2 & 1 & s_2 & & & & \\ 0 & t_3 & 1 & s_3 & & & \\ & & \ddots & \ddots & \ddots & & \\ & & & t_{m-1} & 1 & s_{m-1} & \\ 0 & & & 0 & t_m & 1 & \end{bmatrix} =$$

and that

$$\mathbf{L} := \begin{bmatrix} d_1 & 0 & & & & 0 \\ t_2 & d_2 & 0 & & & \\ 0 & t_3 & d_3 & 0 & & \\ & & \ddots & \ddots & & \\ & & & t_{m-1} & d_{m-1} & 0 \\ 0 & & & & t_m & d_m \end{bmatrix} \tag{4.50}$$

$$\mathbf{R} := \begin{bmatrix} 1 & e_1 & 0 & & & 0 \\ 0 & 1 & e_2 & 0 & & \\ 0 & 0 & 1 & e_3 & & \\ & & & \ddots & \ddots & \\ & & & & 1 & e_{m-1} \\ 0 & & & & 0 & 1 \end{bmatrix} . \tag{4.51}$$

Using the sequential LR decomposition we then have

$$(\mathbf{I} - \mathbf{A}_L\, \mathbf{J})\mathbf{J} = \mathbf{A}_R$$

from (4.46), and

$$\mathbf{I} - \mathbf{A}_L\, \mathbf{J} = \mathbf{D}$$

from (4.45).

In terms of components, taking $s_m := t_1 := 0$, this gives

$$d_1 := 1, e_1 := s_1, d_k := 1 - t_k e_{k-1} \quad (k=2, 3, \ldots, m)$$

$$e_k := s_k/d_k \quad (k=2, 3, \ldots, m-1).$$

The sequential Gauss algorithm

In this case the solution of $\mathbf{A}\,\mathbf{u} = \mathbf{v}$ is obtained from (4.46)–(4.48), namely

$$\begin{aligned} (\mathbf{I} - \mathbf{A}_L\, \mathbf{J})\mathbf{J} &= \mathbf{A}_R \\ (\mathbf{I} + \mathbf{A}_L(\mathbf{I} - \mathbf{J}))\, \mathbf{f} &= \mathbf{v} \\ (\mathbf{I} + \mathbf{J})\mathbf{u} &= \mathbf{f} \ . \end{aligned} \tag{4.52}$$

In components we have

$$e_1 := s_1 \ ;$$

$$e_k := s_k/(1 - t_k e_{k-1}) \qquad k=2, 3, \ldots, m-1;$$

$$f_1 := v_1$$

$$f_k := (v_k - t_k f_{k-1})/(1 - t_k e_{k-1}) \qquad k=2, 3, \ldots, m\,;$$

$$u_m := f_m$$

$$u_k = f_k - e_k u_{k+1}\ , \qquad k=m-1, m-2, \ldots, 1 \ .$$

Both the LR and Gauss algorithms are obviously sequential and so we now define parallel versions according to the principle of vector iteration.

The parallel LR-algorithm

The serial algorithm is now converted to a parallel iterative form by the principle of vector iteration. Suppose that $\mathbf{J}^{(0)}$ is given. Then $\mathbf{J}$ is obtained from (4.46) iteratively as follows:

$$(\mathbf{I} - \mathbf{A}_L \, \mathbf{J}^{(i-1)}) \, \mathbf{J}^{(i)} = \mathbf{A}_R \; , \qquad i=1, 2, \ldots, M \; .$$

M denotes the iteration number at which no change in the values to the accuracy required is obtained by going to an $(M+1)$th iteration. $\mathbf{J}^{(M)}$ having been determined, $\mathbf{D}$ is then calculated from (4.45) in the form

$$\mathbf{D} = \mathbf{I} - \mathbf{A}_L \, \mathbf{J}^{(M)} \; . \tag{4.53}$$

If the non-zero elements of $\mathbf{J}^{(i)}$ are written $e_1^{(i)}, e_2^{(i)}, \ldots, e_{m-1}^{(i)}$ we have in terms of components:

$$e_k^{(0)} \text{ is given for } k=2, 3, 4, \ldots, m-1,$$

and then

$$\begin{aligned} e_1^{(i)} &:= s_1 \, , & i&=0, 1, 2, \ldots, M, \\ e_k^{(i)} &:= s_k \Big/ \left(1 - t_k e_{k-1}^{(i-1)}\right) & i&=1, 2, \ldots, M \, ; \\ & & k&=2, 3, \ldots, m-1 \; . \end{aligned} \tag{4.54}$$

Finally

$$\begin{aligned} d_1 &:= 1 \\ d_k &:= 1 - t_k e_{k-1}^{(M)} \, , \qquad k=2, 3, \ldots, m \; . \end{aligned}$$

The parallel Gauss algorithm

Given $\mathbf{J}^{(0)}$, $\mathbf{f}^{(0)}$ and $\mathbf{u}^{(0)}$ the problem is to solve the system $\mathbf{A}\,\mathbf{u} = \mathbf{v}$ where $\mathbf{A}$ is a tridiagonal matrix with unit diagonal terms. $\mathbf{J}$, $\mathbf{f}$ and $\mathbf{u}$ are obtained iteratively from equations (4.53), (4.47) and (4.48) in the form:

$$\begin{aligned} (\mathbf{I} - \mathbf{A}_L \mathbf{J}^{(i-1)}) \, \mathbf{J}^{(i)} &= \mathbf{A}_R \; , & &(i=1, 2, \ldots, M); \\ (\mathbf{I} - \mathbf{A}_L \mathbf{J}^{(M)}) \, \mathbf{f}^{(i)} &= \mathbf{v} - \mathbf{A}_L \, \mathbf{f}^{(i-1)} \; , & &(i=1, 2, \ldots, N); \\ \mathbf{u}^{(i)} &= \mathbf{f}^{(N)} - \mathbf{J}^{(M)} \, \mathbf{u}^{(i-1)} \; , & &(i=1, 2, \ldots, P) \; . \end{aligned} \tag{4.55}$$

M, N and P are the numbers of repetitions of the $\mathbf{J}$, $\mathbf{f}$ and $\mathbf{u}$ iterations required to achieve the desired accuracy. It is seen that $\mathbf{J}^{(M)}$ is determined first and then used to obtain $\mathbf{f}^{(N)}$. The two are used finally to give $\mathbf{u}^{(P)}$.

In terms of components, the iterations proceed as follows:

Given $e_k^{(0)}$ $(k=2, 3, \ldots, m-1)$,

$$e_1^{(i)} := s_1, \qquad (i=0, 1, \ldots, M), \tag{4.56}$$

$$e_k^{(i)} := s_k/(1-t_k\, e_{k-1}^{(i-1)}),$$

$$(i=1, 2, \ldots, M;\ k=2, 3, \ldots, m-1).$$

Then, given $f_k^{(0)}$ $(k=2, 3, \ldots, m)$,

$$f_1^{(0)} := v_1, \qquad (i=0, 1, 2, \ldots, N),$$

$$f_k^{(i)} := (v_k - t_k f_k^{(i-1)})/(1-t_k e_{k-1}^{(M)}),$$

$$(i=1, 2, \ldots, N;\ k=2, 3, \ldots, m).$$

Finally, given $u_k^{(0)}$

$$u_m^{(i)} := f_m, \qquad (k=1, 2, \ldots, m-1)$$

$$u_k^{(i)} := f_k^{(N)} - e_k^{(M)}\, u_{k+1}^{(i-1)}, \qquad (i=0, 1, 2, \ldots, P)$$

$$(i=1, 2, \ldots, P; k=m-1, m-2, \ldots, 1).$$

Note: Traub [20] has shown that the LR iteration and the Gauss iteration have the effect of reducing the maximum error at each step: that is to say P, M, N can be so chosen that at each step of the f and u iterations the error norm is reduced.

4.1.4 Parallelisation of iterative algorithms

The parallel LR and Gauss algorithms discussed above were obtained by multiplicative decomposition of the matrix $\mathbf{A}$, namely $\mathbf{A} = \mathbf{L}\,\mathbf{R}$. We shall now examine some algorithms based on representing $\mathbf{A}$ as a sum, that is to say

$$\mathbf{A} = \mathbf{A}_L + \mathbf{I} + \mathbf{A}_R.$$

Again $\mathbf{A}$ will be tridiagonal with unit diagonal terms and the notation is the same as in equation (4.43).

We shall look for the parallel elements in the Jacobi, Gauss–Seidel and SOR processes and, capitalising on these, derive parallel versions.

The Jacobi algorithm

The solution of $\mathbf{A}\,\mathbf{u} = \mathbf{v}$ is given by the iteration

$$\mathbf{u}^{(k+1)} = -(\mathbf{A}_L + \mathbf{A}_R)\,\mathbf{u}^{(k)} + \mathbf{v}, \qquad (k=0, 1, 2, \ldots). \tag{4.57}$$

Since this is a matrix vector iteration the procedure is inherently parallel.

The JOR algorithm (Jacobi over-relaxation)
To accelerate the convergence of the Jacobi process a convergence factor $w > 0$ is introduced. Using this we represent the Jacobi procedure as follows:

$$\mathbf{u}^{(k+1)} = ((1-w)\mathbf{I} - w(\mathbf{A}_L + \mathbf{A}_R))\,\mathbf{u}^{(k)} + w\,\mathbf{v}, \qquad (4.58)$$

$$(w > 0 \text{ and } k=0, 1, 2, \ldots).$$

This is obviously parallel too since, as soon as the $\mathbf{u}^{(i)}$ are known, all the components of $\mathbf{u}^{(i+1)}$ can be calculated simultaneously in parallel.

The parallel Gauss–Seidel algorithm
Outer iteration

$$(\mathbf{I} + \mathbf{A}_R)\,\mathbf{u}^{(k)} = -\mathbf{A}_L\mathbf{u}^{(k-1)} + \mathbf{v}\,, \qquad (k = 1, 2, \ldots). \qquad (4.59)$$

By contrast with the Jacobi procedure (4.57) the Gauss–Seidel procedure is not parallel in this form. The specific form of the components arising from (4.59) is

$$u_m^{(k)} = -t_m u_{m-1}^{(k-1)} + v_m$$

$$u_i^{(k)} = -t_i u_{i-1}^{(k-1)} + v_i - e_i\,u_{i+1}^{(k)}, \quad (i=m-1, m-2, \ldots, 2)$$

$$u_1^{(k)} = v_1 - e_1 u_2{}^{(k)} \quad ,$$

showing that the components of $\mathbf{u}^{(k)}$ can only be calculated one after the other and not simultaneously. If, however, we substitute

$$\mathbf{z} := -\mathbf{A}_L\,\mathbf{u}^{(k-1)} + \mathbf{v}, \qquad \mathbf{y} := \mathbf{u}^{(k)}$$

then the components of $\mathbf{z}$ can be calculated in parallel and (4.59) can be written

$$(\mathbf{I} + \mathbf{A}_R)\mathbf{y} = \mathbf{z} \quad . \qquad (4.59')$$

This *bidiagonal* system is solved by the following parallel iteration.

Inner iteration

$$\mathbf{y}^{(i)} = \mathbf{z} - \mathbf{A}_R\,\mathbf{y}^{(i-1)}, \qquad (i=1, 2, 3, \ldots) \qquad (4.60)$$

This is the Jacobi iterative scheme for the solution of the bidiagonal system (4.59′).

Just as the attempt is made to speed up the convergence of the Jacobi procedure by introducing a factor w so the same device can now be applied indirectly to the Gauss–Seidel algorithm. This gives the parallel SOR procedure.

Parallel SOR algorithm
Outer iteration
Corresponding to (4.59) we write

$$(I+w\,\mathbf{A}_R)\,\mathbf{u}^{(k)} = ((1-w)\mathbf{I} - w\,\mathbf{A}_L)\,\mathbf{u}^{(k-1)} + w\,\mathbf{v}, \qquad (4.61)$$

$$(w > 0,\, k=1, 2, \ldots).$$

This converges for $0 < w < 2$.

Now we substitute

$$\mathbf{z}_w := ((1-w)\mathbf{I} - w\,\mathbf{A}_L)\,\mathbf{u}^{(k-1)} + w\,\mathbf{v},$$

$$\mathbf{y}_w := \mathbf{u}^{(k)},$$

and (4.61) is reduced to the form

$$(\mathbf{I} + w\,\mathbf{A}_R)\mathbf{y} = \mathbf{z} \quad . \qquad (4.61')$$

This bidiagonal system is again solved by the following parallel iteration.

Inner iteration

$$\mathbf{y}_w^{(i)} = \mathbf{z}_w - w\,\mathbf{A}_R\,y_w^{(i-1)} \quad , \qquad (i=1, 2, \ldots). \qquad (4.62)$$

4.1.5 Comparison of performance

We compare these processes by using a two-parameter model problem with

$$\mathbf{A}_{s,m} := \begin{bmatrix} 1 & s & & & & 0 \\ s & 1 & s & & & \\ 0 & s & 1 & s & & \\ & & \ddots & \ddots & \ddots & \\ & & & s & 1 & s \\ 0 & & & & s & 1 \end{bmatrix} ,$$

an $m \times m$ matrix of real numbers, with $0 < s < 0.5$.

Comparison of optimal JOR and Jacobi

Let $\rho(\mathbf{C})$ be the spectral radius of a matrix $\mathbf{C}$† and $\mathbf{J}_w$ the iteration matrix for the JOR process, namely

$$\mathbf{J}_w = (1-w)\,\mathbf{I} - w(\mathbf{A}_L + \mathbf{A}_R), \qquad (4.63)$$

(see equation (4.58)).

† The spectral radius is the eigenvalue with the largest absolute value.

It can be shown (see, for example, Stoer, Bulirsch [21]) that the rate of convergence of the JOR procedure improves as $\rho(\mathbf{J}_w)$ decreases and, indeed, a theorem of this type holds in general for iterative processes. We have

Theorem (4.64)
For the model problem

$$\rho(\mathbf{J}_w) \geqslant \rho(\mathbf{J}_1) \ , \qquad w > 0.$$

It then follows that for the model problem the best convergence rate is obtained for $w=1$, that is using the unmodified Jacobi procedure.

Sketch of proof of Theorem (4.64)
We find the eigenvalues of $\mathbf{J}_w$ explicitly for $w > 0$. These are the values of x for which the determinant D_m vanishes, where

$$D_m = \begin{vmatrix} 1-w-x & -ws & 0 & & & 0 \\ -ws & 1-w-x & -ws & 0 & & \\ 0 & -ws & 1-w-x & -ws & & \\ & & \ddots & \ddots & \ddots & \\ & & & & 1-w-x & -ws \\ & 0 & & & -ws & 1-w-x \end{vmatrix}$$

It is easily seen that D_m satisfies the difference equation

$$D_{m+2} - (1-w-x)\,D_{m+1} + w^2 s^2 D_m = 0,$$

and can therefore be represented by

$$D_m = \alpha t_1^m + \beta t_2^m \ ,$$

where α, β are independent of m, and t_1, t_2 satisfy

$$t + w^2 s^2/t = 1-w-x \ .$$

If we put $t = wse^{i\theta}$ we have the relation

$$x = 1 - w - 2ws\cos\theta$$

between x and θ, and an alternative representation

$$D_m = (ws)^m \ [\alpha'\cos m\theta + \beta'\sin m\theta]$$

of D_m . α',β' are determined from the easily verifiable expressions

$$D_1 = 1-w-x = 2ws\cos\theta$$

$$D_2 = (1-w-x)^2 - w^2s^2 = w^2s^2\ (4\cos^2\theta - 1).$$

Solving for α' and β' we get

$$D_m = (ws)^m\ (\cos m\theta + \cot\theta\ \sin m\theta)\ .$$

The eigenvalues are obtained from the set of values of θ for which D_m=0. This entails solving

$$\tan m\theta = -\tan\theta = \tan(\pi-\theta)\ ,$$

and the m values of θ are

$$\theta_i = i\,\pi/(m+1), \qquad i=1, 2, \ldots m.$$

The corresponding eigenvalues x_i are thus

$$x_i = 1-w-2ws\cos\theta_i \quad . \tag{4.65}$$

When $0 < w < 1$ we get

$$\max x_i = x_m = 1-w+2ws\,a,$$

where

$$a = \cos(\pi/(m+1))\ .$$

Thus, comparing with the largest eigenvalue $2sa$ for w=1 we have

$$1-w+2ws\,a = (1-w)\,(1-2as)$$

and this is plainly positive as long as $2as < 1$. Thus the Theorem holds for $0<w<1$.

For $w > 1$ we look at

$$-x_i = w-1 + 2ws\cos\theta_i\ .$$

The spectral radius in this case corresponds to the largest $-x_i$, namely

$$-x_1 = w-1 + 2ws\,a\ .$$

Comparison with $2sa$ gives

$$-x_1 - 2sa = (w-1)\,(2sa+1)\ ,$$

again seen to be positive.

This proves Theorem (4.64).

Since it is now established that the best convergence for the model problem is obtained with the pure Jacobi process, it remains to compare the following algorithms with each other.

(a) parallel Gauss
(b) Jacobi
(c) parallel Gauss–Seidel
(d) parallel optimal SOR.

In order to produce a table of comparisons the error vectors $\mathbf{h}^{(i)}$ at the ith iteration are introduced. For (a), (c) and (d) the relation

$$\|\mathbf{h}^{(k+1)}\| \leqslant C\, \|\mathbf{h}^{(k)}\|, \qquad (k=0, 1, \ldots) \tag{4.66}$$

holds, where C is a constant.

For (b) we have

$$\mathbf{h}^{(k+1)} = \mathbf{W}\, \mathbf{h}^{(k)}, \qquad (k=0,1,2, \ldots) \tag{4.67}$$

where

$$\mathbf{W} := -(\mathbf{A}_L + \mathbf{A}_R) \quad .$$

We now make the following assumption. After N iterations let the error norm be reduced by a factor 2^{-b}, that is

$$\|\mathbf{h}^{(N)}\| = 2^{-b} \|\mathbf{h}^{(0)}\| \quad . \tag{4.68}$$

By (4.66) we have

$$\|\mathbf{h}^{(N+1)}\| \leqslant C^N \|\mathbf{h}^{(0)}\| \quad ,$$

and so

$$C^N \doteqdot 2^{-b},$$

and

$$N \doteqdot -b/\log_2 C \quad .$$

To estimate the error of the Jacobi procedure we have (see Stoer, Bulirsch [21])

$$\lim_{N\to\infty} \sup \left(\frac{\|\mathbf{h}^{(N)}\|}{\|\mathbf{h}^{(0)}\|} \right)^{1/N} \leqslant \rho\,(\mathbf{J}) \quad .$$

To satisfy (4.68) the number N of iterations is calculated approximately to be

$$N = -b/\log_2 \rho\,(\mathbf{J}) \quad .$$

Using these estimates Traub [20] gives estimates for the computer time needed for the model problem shown in Table 4.1 in which the notation used is:

$$\|\mathbf{h}^{(N)}\| = 2^{-b}\|h^{(0)}\| \qquad \text{(requirement)}$$

$$\epsilon^2 = 1-4s^2$$

$$\sigma^2 = \epsilon^2 + (\pi/(m+1))^2$$

$$\mu = -\log_2 s \; .$$

Table 4.1. Overall time of each algorithm

	$s \doteqdot ½$	s small	$s \doteqdot ¼$
Parallel Gauss	$10b/3\epsilon$	$5b/\mu$	$5b/2$
Jacobi	$4b/\sigma^2$	$3b/(\mu-1)$	$3b$
Parallel Gauss–Seidel	$4b^2/3\sigma^2$	$b^2/(\mu(\mu-1))$	$b^2/2$
Parallel optimal SOR	$4b^2/9\sigma^2$	b^2/μ^2	$b^2/4$

Conclusions

1. Parallel SOR is superior to parallel Gauss–Seidel.
2. Jacobi and parallel Gauss are superior to the other two.
3. For small s (i.e. $\mathbf{A}$ is strongly diagonal dominant) Jacobi is superior to parallel Gauss.
4. For moderate s (¼) Jacobi and parallel Gauss are about the same.
5. For s near to ½ (very small diagonal dominance) parallel Gauss is considerably superior to Jacobi.

The boundary lies at about $s = 0.19$: that is, for $s \leqslant 0.19$ the Jacobi algorithm is the best of those considered, while for $s > 0.19$ parallel Gauss is the best.

4.2 Treatment of the eigenvalue problem

First we shall review briefly the processes to be described for the solution of the eigenvalue problem

$$\mathbf{A}\,\mathbf{x} = \lambda\,\mathbf{x} \qquad (\mathbf{x} \neq 0),$$

where $\mathbf{A}$ is an $n \times n$ matrix of real components a_{ik}, while λ, and the elements of the n vector $\mathbf{x}$ are real also. Although attention is confined to real matrices, generalisation to complex matrices is possible for all the procedures described.

In practice most processes used for the determination of eigenvalues use similarity transformations of the matrix $\mathbf{A}$. These transformations leave the eigenvalues and eigenvectors unaltered. For let $\mathbf{T}$ be a nonsingular matrix. Then the following three forms are equivalent:

$$(\mathbf{T}^{-1}\mathbf{A}\,\mathbf{T})\,(\mathbf{T}^{-1}\,\mathbf{x}) = \lambda\,(\mathbf{T}^{-1}\,\mathbf{x})\ ;$$

$$\mathbf{T}^{-1}\,\mathbf{A}\,\mathbf{x} = \lambda\,\mathbf{T}^{-1}\,\mathbf{x}\ ;$$

$$\mathbf{A}\,\mathbf{x} = \lambda\mathbf{x}\ .$$

If now we construct a series of similar matrices

$$\mathbf{A}_0 := \mathbf{A},$$

$$\mathbf{A}_{k+1} := \mathbf{T}_k^{-1}\,\mathbf{A}_k\,\mathbf{T}_k \qquad (k=0, 1, 2, \ldots)$$

then all $\mathbf{A}_k$ have the same eigenvalues. If the $\mathbf{T}_k$ are such that the sequence $\mathbf{A}_k$ tends to a diagonal matrix, the eigenvalues are the elements of the diagonal and can be read off directly from it.

First we shall investigate the Jacobi procedure for symmetric matrices. This is based on use of the rotation matrices $\mathbf{U}_{pq}$ shown below for the transformation.

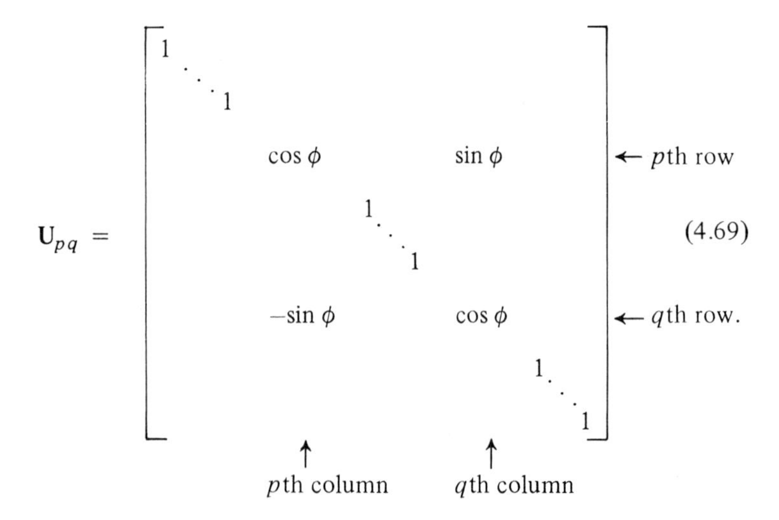

It is easily verified that these $\mathbf{U}$ matrices are orthogonal, that is $\mathbf{U}^{\mathrm{T}} = \mathbf{U}^{-1}$, a property usually required of transformation matrices. The angle ϕ is so chosen that the elements $a'_{pq} = a'_{qp}$ of the transformed matrix $\mathbf{A}' = \mathbf{U}\,\mathbf{A}\,\mathbf{U}^{\mathrm{T}}$ vanish.

Although the Jacobi process is stable it is slow to converge. A process to be preferred is the so-called QR process in which the transformations are so chosen that $\mathbf{U}^{\mathrm{T}}\mathbf{A} = \mathbf{R}$ is an upper triangular matrix. For reasons of economy of calculation it is supposed that **A** is either of upper Hessenberg form, or, in the symmetrical case, tridiagonal. For this reason a preliminary discussion is given of the parallelisation of the Householder procedure for the reduction of a symmetrical matrix to tridiagonal form. This leads finally to a special form of the QR process for tridiagonal matrices. For this algorithm the results about linear recurrences developed in 3.3 are used. Kuck and Sameh [22], [23] and Sameh [24] have investigated these algorithms.

In addition the Hyman procedure is explained. This uses the characteristic polynomial. Ward [25] has shown that it is possibly preferable on a pipeline computer to the QR procedure. Finally a procedure involving simultaneous iteration is considered. This has been implemented by Schendel, Gomm and Weistroffer [26] on a TR 440 using simulation routines for a vector processor.

4.2.1 The Jacobi procedure

Sameh [24], and Kuck and Sameh [22], have investigated the implementation of the Jacobi process on an ILLIAC IV.

The basic idea for this parallelisation lies in the choice made of the angle ϕ of the transformation matrix $\mathbf{U}_{pq}$ (see(4.69)). It is desirable to be able to calculate several such angles in parallel. In fact, this creates no difficulty when to each angle correspond different row (column) subscripts. Then several transformations can be effected simultaneously.

Consider the following transformation matrix **R** for application to a 4X4 symmetrical matrix **A**,

$$\mathbf{R} = \mathbf{U}_{12}\,\mathbf{U}_{34} \;.$$

This gives the specific form

$$R = \begin{bmatrix} c_1 & s_1 & 0 & 0 \\ -s_1 & c_1 & 0 & 0 \\ 0 & 0 & c_2 & s_2 \\ 0 & 0 & -s_2 & c_2 \end{bmatrix} , \qquad (4.70)$$

where $c_1 = \cos\phi_{12}$, $s_1 = \sin\phi_{12}$, $c_2 = \cos\phi_{34}$ $s_2 = \sin\phi_{34}$. When applied to the symmetrical 4X4 real matrix **A** the angles ϕ_{12} and ϕ_{34} can be calculated in parallel to reduce to zero the terms $a'_{12} = a'_{21}$, $a'_{34} = a'_{43}$ in the transformed matrix $\mathbf{A}' = \mathbf{R}\,\mathbf{A}\,\mathbf{R}^{\mathrm{T}}$. It is easily checked specifically that

$$a'_{12} = a_{12}\cos 2\phi_{12} - \tfrac{1}{2}(a_{11} - a_{22})\sin 2\phi_{12} ,$$

$$a'_{34} = a_{34}\cos 2\phi_{34} - \tfrac{1}{2}(a_{33} - a_{44})\sin 2\phi_{34}$$

so that

$$\tan 2\phi_{12} = \frac{2a_{12}}{a_{11} - a_{22}} \;, \quad \tan 2\phi_{34} = \frac{2a_{34}}{a_{33} - a_{44}} \quad .$$

This completely defines the transformation matrix **R** and hence **A**$'$. Two of the 6 superdiagonal terms have now been reduced to zero and the Jacobi method is to proceed (in this 4X4 case) to carry out two further transformations on **A**$'$ in such a way as to reduce to zero two other of the so far untreated superdiagonal terms. Thus we could next take a new transformation matrix

$$\mathbf{R} = \mathbf{U}_{13}\,\mathbf{U}_{24} = \begin{bmatrix} c_1 & 0 & s_1 & 0 \\ 0 & c_2 & 0 & s_2 \\ -s_1 & 0 & c_1 & 0 \\ 0 & -s_2 & 0 & c_2 \end{bmatrix} \;,$$

where now

$$c_1, s_1 = \begin{matrix}\cos \\ \sin\end{matrix}\,\phi_{13}, \quad c_2, s_2 = \begin{matrix}\cos \\ \sin\end{matrix}\,\phi_{24} \quad .$$

Then we would construct transformation $\mathbf{A}'' = \mathbf{R}\,\mathbf{A}'\,\mathbf{R}^{\mathrm{T}}$, choosing ϕ_{13} and ϕ_{24} so as to zeroise the terms $a''_{13} = a''_{31}$, $a''_{24} = a''_{42}$. Finally, the so-called sweep is completed by applying to **A**$''$ the transformation

$$\mathbf{R} = \mathbf{U}_{14}\,\mathbf{U}_{23} = \begin{bmatrix} c_1 & 0 & 0 & s_1 \\ 0 & c_2 & s_2 & 0 \\ 0 & -s_2 & c_2 & 0 \\ -s_1 & 0 & 0 & c_1 \end{bmatrix} \;,$$

where

$$\begin{matrix}c_1 \\ s_1\end{matrix} = \begin{matrix}\cos \\ \sin\end{matrix}\,\phi_{14} \qquad \begin{matrix}c_2 \\ s_2\end{matrix} = \begin{matrix}\cos \\ \sin\end{matrix}\,\phi_{23} \;,$$

and these angles are so chosen as to zeroise the terms $a'''_{14} = a'''_{41}$, $a'''_{23} = a'''_{32}$ in $\mathbf{A}''' = \mathbf{R}\,\mathbf{A}''\,\mathbf{R}^{\mathrm{T}}$.

This succession of transformations is such as to leave a matrix $\mathbf{A}_1 = \mathbf{A}'''$ with the same eigenvalues as the original **A** and, in principle, a diagonal which, owing to the successive rotational transformations, is relatively more prominent with respect to the remaining elements.

Further sweeps are conducted starting with $\mathbf{A}_1$, and the succeeding iterations $\mathbf{A}_2, \mathbf{A}_3$ should tend more and more to a purely diagonal form.

If we generalise this discussion to a symmetric $n \times n$ matrix $\mathbf{A}$ we see that in each of the corresponding $n \times n$ transformations $\mathbf{R}$ of a sweep, $[n/2]$ angles can be calculated in parallel so as to zeroise $[n/2]$ superdiagonal elements. This parallel calculation of the angles is the first requirement of a parallel form of the algorithm.

Now there are $(n^2-n)/2$ superdiagonal terms and if at each step of a sweep $n/2$ are reduced to zero, then, in order to reduce all such elements to zero once each during the sweep, as required by the Jacobi procedure, a sweep must consist of the following number of steps:

$$\frac{n^2-n}{2} \Big/ \frac{n}{2} = \begin{cases} n-1 & \text{(even } n) \\ n & \text{(odd } n) \end{cases} .$$

This is the second requirement demanded of a parallel version of the Jacobi algorithm.

Both requirements can be met as follows.

$\mathbf{A} = (a_{ij}\colon i,j = 0, 1, \ldots, n-1)$ is a symmetric real matrix and n is taken as even. The transformation matrix used at each step is always

$$\mathbf{U} = \operatorname{diag}(\mathbf{U}_0, \mathbf{U}_1, \ldots, \mathbf{U}_{(n/2)-1}) \tag{4.71}$$

where

$$\mathbf{U}_k = \begin{bmatrix} \cos\phi_k, & \sin\phi_k \\ -\sin\phi_k, & \cos\phi_k \end{bmatrix} \qquad (k=0, 1, \ldots, (n/2)-1)$$

and the angles are so chosen as to reduce to zero the elements $a'_{2k,2k+1}$ next to the diagonal in the transformed matrix $\mathbf{A}' = \mathbf{U}\,\mathbf{A}\,\mathbf{U}^{\mathrm{T}}$.

The transformation process at each step can be viewed in terms of a partitioning of $\mathbf{A}$ into 2×2 submatrices

$$\mathbf{A}_{jk} = \begin{bmatrix} a_{2j,2k} & a_{2j,2k+1} \\ a_{2j+1,2k} & a_{2j+1,2k+1} \end{bmatrix}$$

$$(j,\, k=0, 1, \ldots, (n/2)-1, j \leqslant k \text{ by symmetry}).$$

Then $\mathbf{A}'$ is composed of 2×2 submatrices

$$\mathbf{A}'_{jk} = \mathbf{U}_j \mathbf{A}_{jk} \mathbf{U}_k^{\mathrm{T}} \qquad (j,\, k=0, 1, \ldots, (n/2)-1, j \leqslant k) .$$

In order to satisfy the second requirement concerning the zeroising of the elements a slightly different procedure is adopted from that described above which entailed permutations of the elements of the transformation matrix. Instead, the elements of the $\mathbf{A}'$ matrices are subjected to a permutation transformation at the end of each of the $n-1$ steps of a sweep and the same transformation $\mathbf{U}$ of (4.71) applied with the fixed choice of angles. Thus having obtained $\mathbf{A}'$ at the end of a certain step we first calculate

$$\mathbf{B} = \psi\, \mathbf{A}'\, \psi^{\mathrm{T}}$$

and then a new

$$\mathbf{A}' = \mathbf{U}\,\mathbf{B}\,\mathbf{U}^{\mathrm{T}} \; .$$

For the first $n-2$ steps of each sweep we use

$$\psi := (\mathbf{i}_1\,;\, \mathbf{i}_n\,;\, \mathbf{i}_2\,;\, \ldots\,;\, \mathbf{i}_{n-1}) = \begin{bmatrix} 1 & 0 & 0 & & & \\ 0 & 0 & 1 & & & \\ 0 & 0 & 0 & 1 & & \\ \vdots & \vdots & \vdots & & \ddots & 1 \\ 0 & 1 & 0 & & & 0 \end{bmatrix} , \tag{4.72}$$

where $\mathbf{i}_j$ is the jth column of the standard unit $n \times n$ matrix. For the final $(n-1)$th step of a sweep we use

$$\psi' := (\mathbf{i}_n; \mathbf{i}_1; \ldots; \mathbf{i}_{n-1}) = \begin{bmatrix} 0 & 1 & 0 & & \\ 0 & 0 & 1 & & \\ 0 & 0 & 0 & \ddots & \\ \vdots & \vdots & \vdots & \ddots & 1 \\ 1 & 0 & 0 & & 0 \end{bmatrix} . \tag{4.73}$$

Sweeps are repeated until the matrix $\mathbf{A}$ has been reduced to a prescribed accuracy to diagonal form.

As an example, the 4×4 matrix

$$\mathbf{A}_0 := \begin{bmatrix} 1, & -1, & -2 & 2 \\ -1 & 2 & 1 & -1 \\ -2 & 1 & 3 & 2 \\ 2 & -1 & 2 & 4 \end{bmatrix}$$

has the following four eigenvalues:

$$5.69022935,\, 4.88944325,\, -1.62149407,\, 1.04182150 \; .$$

It turns out that three sweeps with the algorithm suffice. The iterates $\mathbf{A}_1, \mathbf{A}_2, \mathbf{A}_3$ at the end of the sweeps are:

$$\mathbf{A}_1 = \begin{bmatrix} 5.59781973 & 0.24788418 & -0.25478364 & 0 \\ & 0.48163658 & 0 & 1.12731811 \\ & & 4.74289490 & 1.11056566 \\ & & & -0.82235121 \end{bmatrix}$$

$$\mathbf{A}_2 = \begin{bmatrix} -1.62069887 & 0 & -0.07166821 & 0 \\ & 1.04187490 & 0 & -0.01664738 \\ & & 4.88865949 & -0.00204275 \\ & & & 5.69016448 \end{bmatrix}$$

$$\mathbf{A}_3 = \begin{bmatrix} 5.69022931 & 0 & -0.00000175 & 0 \\ & 1.04182150 & 0 & -0.00000093 \\ & & 4.88944325 & 0 \\ & & & -1.62149406 \end{bmatrix}$$

The lower diagonal terms can be filled in by symmetry. 0 means less than 1×10^{-8}.

Kuck and Sameh [22] have investigated the implementation of the procedure for a 16X16 matrix using $p=16$ processors. The main problem is to develop a suitable organisation for the data in store. Figure 4.1 shows the scheme at the beginning of the procedure.

LOC	ROW	PE 0	1	2	3	4	5	6	7	8	9	10	11	12	13	14	15
0	0	0 0	0 4	0 8	0 12	0 1	0 5	0 9	0 13	0 2	0 6	0 10	0 14	0 3	0 7	0 11	0 15
1	1	1 14	1 3	1 7	1 11	1 15	8 8	1 4	1 8	1 12	1 1	1 5	1 9	1 13	1 2	1 6	1 10
2	2	2 9	2 13	2 2	2 6	2 10	2 14	2 3	2 7	2 11	2 15	8 9	2 4	2 8	2 12	9 9	2 5
3	3	3 4	3 8	3 12	9 10	3 5	3 9	3 13	10 10	3 6	3 10	3 14	3 3	3 7	3 11	3 15	8 10
4	4	11 11	4 7	4 11	4 15	8 11	4 4	4 8	4 12	9 11	4 5	4 9	4 13	10 11	4 6	4 10	4 14
5	5	5 13	10 12	5 6	5 10	5 14	11 12	5 7	5 11	5 15	8 12	12 12	5 8	5 12	9 12	5 5	5 9
6	6	6 8	6 12	9 13	13 13	6 9	6 13	10 13	6 6	6 10	6 14	11 13	6 7	6 11	6 15	8 13	2 13
7	7	7 7	7 11	7 15	8 14	12 14	7 8	7 12	9 14	13 14	7 9	7 13	10 14	14 14	7 10	7 14	11 14
8	8	10 15	14 15			11 15	15 15			8 15	12 15			9 15	13 15		

Fig. 4.1 – Data storage for matrix of 16X16 elements.

The entries are the subscripts (i, j) of the element a_{ij}. For instance $a_{2,11}$ is stored in location 2 of processor 8.

The following formulae show how the allocation is determined.

(i) $0 \leqslant i \leqslant \frac{1}{2}p-1$. a_{ij} is placed in location i of processing element (processor) number

$$[\sqrt{p}(j(\text{mod}\sqrt{p})) + j/\sqrt{p} + i(\sqrt{p}+1)] \qquad (\text{mod}\, p).$$

(ii) $p/2 \leqslant i \leqslant p-1$. a_{ij} is placed in location $j-\frac{1}{2}p+1$ of processing element number

$$[\sqrt{p}(i(\text{mod}\sqrt{p}) + 1) + \frac{(i-\frac{1}{2}p)}{\sqrt{p}} + (j-\frac{1}{2}p)(\sqrt{p}+1)+1] \ (\text{mod}\, p). \tag{4.74}$$

In determining the time required by this procedure, the time required to compute the $\frac{1}{2}n$ angles and to effect the permutation transformations at each step may be neglected. Because of the adroit choice made for data storage it is necessary in general to shift only 3 rows and $\sqrt{p}-1$ columns. The most onerous task is the transformation

$$\mathbf{A}' = \mathbf{U}\,\mathbf{A}\,\mathbf{U}^{\mathrm{T}} \ .$$

With p processors this can be implemented most satisfactorily on $\sqrt{p} \times \sqrt{p}$ submatrices. If $\sqrt{p}$ is even this gives no difficulty as we saw in the earlier description of the 4×4 example. An inefficiency arises only in the transformation of the diagonal blocks since in this respect in the sequential case symmetry would be respected and the elements to be annihilated would be set directly to zero. However, the loss in efficiency due to this step and the overall degradation inflicted on the whole procedure is easily calculated as will now be shown.

Let $\mathbf{A}$ be an $n \times n$ matrix in which $n=l\sqrt{p}$, where p is the number of processors available and l is an integer. The number of superdiagonal blocks is $\sum_{i=1}^{l-1} i = \frac{1}{2}l(l-1)$ each of dimensions $\sqrt{p} \times \sqrt{p}$. In addition, the diagonal has l such blocks for which, in the sequential case, only half the computational effort is needed. Thus, the efficiency is given by

$$E_p = \frac{l(l-1)/2 + l/2}{l(l-1)/2 + l} = \frac{l}{l+1} \tag{4.75}$$

When $p=n=64$ ($l=8$) the efficiency is about 89%.

Eberlein and Boothroyd [27] proposed a 'Jacobi-like' algorithm for an unsymmetrical matrix $\mathbf{A}$. This can also be parallelised in a similar fashion. Sameh [24] gives a penetrating investigation of the procedure.

4.2.2 The Householder procedure

The Householder algorithm reduces a symmetric matrix $\mathbf{A} = [a_{ij}\colon i,j=0, 1, \ldots, n-1]$ to tridiagonal form. Calculation of the eigenvalues can then proceed, for example by use of the QR algorithm described in section 4.2.3.

The Householder method entails using an $n \times n$ matrix transformation $\mathbf{T}$ of the following form

$$\mathbf{T} = \mathbf{I} - \beta\,\mathbf{u}\,\mathbf{u}^{\mathrm{T}} \tag{4.76}$$

where $\mathbf{u}$ is an n-vector of real terms, β is a real scalar coefficient and $\mathbf{u}^T\mathbf{u} = 1$. This will be described in detail below.

The reduction takes place in $n-2$ steps. Just before the kth step the original matrix $\mathbf{A}$ will have the form

$$\mathbf{A}_{k-1} = \left[\begin{array}{cccc|c|c} \delta_0 & \gamma_1 & & & & \\ \gamma_1 & \ddots & \ddots & \gamma_{k-1} & & \\ & \ddots & \gamma_{k-1} & \delta_{k-1} & \gamma_k & \\ \hline & & & \gamma_k & \delta_k & \mathbf{a}_k^T \\ \hline & & \mathbf{0} & & \mathbf{a}_k & \hat{\mathbf{A}}_{k-1} \end{array}\right] \tag{4.77}$$

the kth row $\mathbf{a}_k^T$ and column $\mathbf{a}_k$ being shown explicitly. In the kth step the following elements are annihilated:

(i) All those in the kth column below the lower neighbouring diagonal;
(ii) All those in the kth row to the right of the upper neighbouring diagonal.

Symmetry is preserved and, in particular, terms which have been reduced to zero are unaffected by the remaining transformations.

The precise form of the transformation $\mathbf{T}_k$ to be applied at the kth ($k=1, 2, \ldots, n-2$) step is

$$\mathbf{T}_k = \mathbf{I} - \beta_k \mathbf{u}_k \mathbf{u}_k^T \tag{4.78}$$

where

$$\mathbf{u}_k^T = (0, 0, \ldots, 0, a_{k-1,k} \pm S_k, a_{k-1,\,k+1}, \ldots, a_{k-1,\,n-1})$$

$$S_k = + \left\{ \sum_{i=k}^{n-1} a_{k-1,i}^2 \right\}^{½} \tag{4.79}$$

$$\beta_k = [S_k(S_k \pm a_{k-1,\,k})]^{-1} \quad .$$

The sign ambiguity is resolved by the choice

$$|a_{k-1,\,k} \pm S_k| = |a_{k-1k}| + S_k \quad . \tag{4.80}$$

$\mathbf{T}_k$ is symmetrical.

By symmetry the transformation

$$\mathbf{A}_k = \mathbf{T}_k \mathbf{A}_{k-1} \mathbf{T}_k \qquad (k=1, 2, \ldots, n-2) \tag{4.81}$$

can be rewritten as follows:

$$\mathbf{A}_k = \mathbf{A}_{k-1} - \mathbf{q}_k \mathbf{u}_k^{\mathrm{T}} - \mathbf{u}_k \mathbf{p}_k^{\mathrm{T}} \tag{4.82}$$

where

$$\mathbf{q}_k = \mathbf{p}_k - \beta_k (\mathbf{u}_k \mathbf{u}_k^{\mathrm{T}} \mathbf{p}_k) \tag{4.83}$$

$$\mathbf{p}_k = \beta_k \mathbf{A}_{k-1} \mathbf{u}_k \quad .$$

As an example we demonstrate the two steps necessary to tridiagonalise the 4×4 symmetrical matrix

$$\mathbf{A}_0 = \begin{bmatrix} 1 & -1 & -2 & 2 \\ -1 & 2 & 1 & -1 \\ -2 & 1 & 3 & 2 \\ 2 & -1 & 2 & 4 \end{bmatrix}$$

whose eigenvalues were computed in Section 4.2.1 as an example of the Jacobi procedure. Using (4.82) and (4.83) twice we get:

$$\begin{bmatrix} 1 & 3 & 0 & 0 \\ 3 & 2.4444 & 1.2222 & -0.5555 \\ 0 & 1.2222 & 3.1111 & 2.2222 \\ 0 & -0.5555 & 2.2222 & 3.4444 \end{bmatrix}$$

and then

$$\begin{bmatrix} 1 & 3 & 0 & 0 \\ 3 & 2.4444 & -1.3426 & 0 \\ 0 & -1.3426 & 1.4939 & -1.3356 \\ 0 & 0 & -1.3356 & 5.0616 \end{bmatrix} \quad .$$

Formulae (4.79), (4.83) and (4.82) constitute the parallel algorithm.

As in the case of the Jacobi process the implementation will be considered in the case of $p=n$ processors. Each step in the transformation can again most conveniently be performed on $\sqrt{p} \times \sqrt{p}$ submatrices, so again $p=n$ will be assumed to be a perfect square. To use more processors would obviously decrease the efficiency of the procedure considerably since the number of non-zero elements in $\mathbf{A}_k$ and $\mathbf{u}_k$ decreases at each step. The calculation of $\mathbf{u}_k$, $\mathbf{p}_k$ and $\mathbf{q}_k$ follows directly from (4.79) and (4.83). The recursive doubling technique is to be used for S_k. The calculation of $\mathbf{A}_k$ from (4.82) is then arranged in terms of submatrices of dimension $\sqrt{p} \times \sqrt{p}$. Correspondingly the vectors $\mathbf{u}_k$ and $\mathbf{q}_k$ should be divided into subvectors with dimension $\sqrt{p}$. (4.82) is then written

$$\mathbf{A}_{ij}^{(k)} = \mathbf{A}_{ij}^{(k-1)} - \mathbf{q}_i^{(k)} \mathbf{u}_j^{(k)\mathrm{T}} - \mathbf{u}_i^{(k)} \mathbf{p}_j^{(k)\mathrm{T}}$$

for $1 \leqslant i \leqslant j \leqslant \sqrt{p}$, where the number of the iterate is shown as a bracketed superscript.

To calculate the efficiency of the procedure we remark first that the number of operations required to carry the process out sequentially is

$$\sum_{j=1}^{n} \sum_{i=1}^{j} i = \tfrac{1}{2} \sum_{j=1}^{n} j(j+1) = \tfrac{1}{6} n(n+1)(n+2) \quad . \tag{4.84}$$

In the parallel version the same operations are carried out on $\sqrt{n} \times \sqrt{n}$ blocks. Then (4.84) becomes

$$\sum_{j=1}^{\sqrt{n}-1} \sum_{i=1}^{j} i + 2 \sum_{i=1}^{\frac{1}{2}\sqrt{n}} i = \tfrac{1}{6} n^{\frac{1}{2}} (n+\tfrac{3}{2}n^{\frac{1}{2}}+2) \quad . \tag{4.85}$$

This must be multiplied by the number of processors p (here $p=n$) and by the number of storage rearrangements which, in the case of the Kuck and Sameh [22] scheme, is about $\sqrt{p}$. Hence we have for the efficiency with $p=n$

$$E_n = \frac{n^2 + 3n+2}{n^2+1.5n^{3/2}+2n} \quad , \tag{4.86}$$

giving about 86% for $p=n=64$, about the same as the Jacobi process (which, however, gets the eigenvalues).

4.2.3 The QR-procedure

General observations

The QR-procedure† entails the construction of a matrix sequence ($\mathbf{A}_k$: k=1, 2, . . .) by transformations

$$\mathbf{A}_{k+1} = \mathbf{Q}_k^{\mathrm{T}} \mathbf{A}_k \mathbf{Q}_k \tag{4.87}$$

where the transformation matrices $\mathbf{Q}_k$ are unitary and orthogonal and formed in such a way that at each step

$$\mathbf{Q}_k^{\mathrm{T}} \mathbf{A}_k = \mathbf{R}_k \tag{4.88}$$

† This procedure is well-known to specialists. A recent reference, strongly recommended for its clarity, and the completeness of its references is Watkins [28].

is an upper triangular matrix. Another way of defining the transformation is to require that $\mathbf{A}_k$ be decomposed,

$$\mathbf{A}_k = \mathbf{Q}_k\mathbf{R}_k \quad , \tag{4.89}$$

into the product of a unitary orthogonal and an upper triangular matrix.

It is well known that such a decomposition exists for nonsingular matrices. The computational demand for general nonsingular matrices is, however, extremely heavy and consequently this description of the QR-procedure will be confined to the case of upper Hessenberg matrices, namely upper triangular matrices with non-zero subdiagonal elements, and zeros below the subdiagonal.

If the eigenvalues are distinct the sequence $\mathbf{A}_k$ converges to an upper triangular matrix with the eigenvalues in the principal diagonal. To accelerate convergence the most widely used device is the so-called 'shifted QR-algorithm' which makes use of a spectral displacement of $\mathbf{A}$. Instead of seeking the decomposition (4.89) for $\mathbf{A}_k$ we consider $A_k - t_k \mathbf{I}$ where t_k is a real number called the shift of origin. The procedure then takes the following form:

$$\mathbf{A}_0 := \mathbf{A}$$

For k=0, 1, . . .

(i) Determine t_k.
(ii) Make the decomposition

$$\mathbf{A}_k - t_k \mathbf{I} = \mathbf{Q}_k \mathbf{R}_k \quad . \tag{4.90}$$

(iii) Define

$$\mathbf{A}_{k+1} := \mathbf{R}_k \mathbf{Q}_k \quad .$$

Then $\mathbf{A}_k$ has the same eigenvalues as $\mathbf{A}_0 - \sum_{i=0}^{k-1} t_i \mathbf{I}$. The shift is usually chosen as the last element in the main diagonal a_{nn}, or one or both of the eigenvalues of the last 2X2 submatrix in the diagonal. When no special assumption about symmetry is made a problem arises from the fact that the eigenvalues may be conjugate complexes. Francis [29] has proposed a 'double shift algorithm' to avoid complex arithmetic. This entails carrying out two steps at each iteration.

In this case (4.90) becomes:

$$\mathbf{A}_0 := \mathbf{A}$$

For k=0, 2, 4, . . .

(i) Form the decomposition

$$\mathbf{B}_k := (\mathbf{A}_k - t_k \mathbf{I})(\mathbf{A}_k - t_{k+1} \mathbf{I}) = \mathbf{W}_k \mathbf{U}_k \quad . \tag{4.91}$$

(ii) Set

$$\mathbf{A}_{k+2} := \mathbf{W}_k^{\mathrm{T}} \mathbf{A}_k \mathbf{W}_k \quad .$$

If the shifts of origin t_k, t_{k+1} are conjugate complex $\mathbf{B}_k$ is real. $\mathbf{W}_k$ is again an orthogonal matrix, and $\mathbf{U}_k$ is an upper triangular matrix. Thus

$$\begin{aligned} \mathbf{W}_k &= \mathbf{Q}_k \mathbf{Q}_{k+1} \\ & \qquad\qquad\qquad (k=0, 2, 4, \ldots) \\ \mathbf{U}_k &= \mathbf{R}_{k+1} \mathbf{R}_k \quad . \end{aligned}$$

4.2.4 The QR-algorithm for an upper Hessenberg matrix

We shall now investigate in detail a transformation step

$$\mathbf{A}' = \mathbf{W}^{\mathrm{T}} \mathbf{A} \mathbf{W} \tag{4.92}$$

of the algorithm (4.91) for an upper Hessenberg matrix $\mathbf{A} = [a_{ij}: i,j=0, 1, \ldots, n-1]$.

The decomposition of $\mathbf{B}_k$ in (4.91) is not unique. Uniqueness is accomplished by fixing the first column of $\mathbf{W}_k$. If this is chosen as the first column of a Householder matrix $\mathbf{P}_0$ which eliminates the elements below the diagonal in the first column of $\mathbf{B}_k$, then $\mathbf{W}_k$ can be represented as a sequence of $n-1$ Householder matrices and (4.92) can be represented as:

$$\begin{aligned} \mathbf{C}_0 &:= \mathbf{A} \\ \mathbf{C}_{r+1} &:= \mathbf{P}_r^{\mathrm{T}} \mathbf{C}_r \mathbf{P}_r, \quad r=0, 1, \ldots, n-2 \\ \mathbf{A}' &:= \mathbf{C}_{n-1} \quad . \end{aligned} \tag{4.93}$$

The matrix $\mathbf{P}_r$ is so chosen as to eliminate the elements below the diagonal in the rth column of $\mathbf{C}_r$, and thus $\mathbf{A}'$ is again of upper Hessenberg form. The $\mathbf{P}_r$ are given by:

$$\begin{aligned} \mathbf{P}_r &:= \mathbf{I} - (2\, \mathbf{y}_r \mathbf{y}_r^{\mathrm{T}} / \|\mathbf{y}_r\|^2), \\ \mathbf{y}_r^{\mathrm{T}} &:= (\underbrace{0, 0, \ldots, 0}_{r}, 1, s_r, t_r, 0, \ldots, 0) \end{aligned} \tag{4.94}$$

$$\begin{aligned} s_r &:= c_{r+1, r-1}^{(r)} / (c_{r, r-1}^{(r)} \pm S_r) \\ t_r &:= c_{r+2, r-1}^{(r)} / (c_{r, r-1}^{(r)} \pm S_r) \quad , \\ S_r &:= \{ (c_{r, r-1}^{(r)})^2 + (c_{r+1, r-1}^{(r)})^2 + (c_{r+2, r-1}^{(r)})^2 \}^{1/2} \quad , \end{aligned} \tag{4.95}$$

for $r=1, 2, \ldots, n-2$.

For $r=0$:

$$s_0 := b_{10}/(b_{00} + S_0)$$

$$t_0 := b_{20}/(b_{00} \pm S_0) \tag{4.96}$$

$$S_0 := \{b_{00}^2 + b_{10}^2\}^{1/2} \quad .$$

The convention concerning the sign ambiguity can be inferred from the description in Section 4.2.2.

The implementation of (4.93) is carried out in a manner similar to that of the Householder procedure, namely

$$\mathbf{C}_{r+1} := \mathbf{C}_r - \mathbf{v}_r\,\mathbf{p}_r^{\mathrm{T}} - (\mathbf{q}_r - \alpha_r\,\mathbf{y}_r)\mathbf{v}_r^{\mathrm{T}} \tag{4.97}$$

where

$$\mathbf{p}_r^{\mathrm{T}} = \mathbf{y}_r^{\mathrm{T}}\,\mathbf{C}_r$$

$$\mathbf{q}_r = \mathbf{C}_r\mathbf{y}_r \qquad (r=0, 1, \ldots, n-2) \tag{4.98}$$

$$\mathbf{v}_r = 2\,\mathbf{y}_r/\|\mathbf{y}_r\|^2$$

$$\alpha_r = \mathbf{p}_r^{\mathrm{T}}\,\mathbf{v}_r \quad .$$

In this form the algorithm is adapted to parallel implementation. A crucial element is the need to calculate a square root in the formula for S_r. In this connection the special procedure for tridiagonal matrices, to be dealt with next, has the advantage of avoiding the square root.

Kuck and Sameh [22] propose an algorithm which, using at most eight processors, needs 20 time units to calculate $\|y_r\|^2$, s_r and t_r. This is in contrast with 35 time units in the sequential case. The speed gain is not considerable in spite of the ingeniousness of the procedure. For example, in the last step, three divisions are executed in parallel although each entails completely different computations. This is shown in Fig. 4.2.

$$x := c_{r,\,r-1} + S_r$$

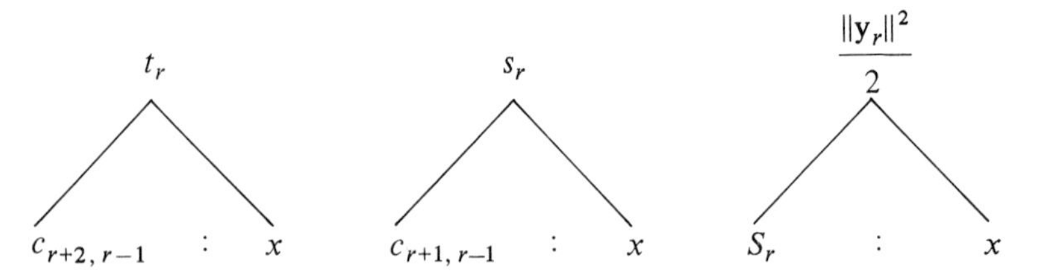

Fig. 4.2

Kuck and Sameh [22] obtain the following expression for the efficiency E_p with p processors and $n=m\sqrt{p}$ $(m \leqslant 1)$

$$E_p = \frac{7.5m + 35/p}{7.5m + 27.5} \quad . \tag{4.99}$$

This gives for $n=256$ and $p=16$ an efficiency measure of 82.5%, again about that of the Jacobi procedure. An increase in p leads to a rapid decrease in E_p: for example, with $n=p=256$, E_p is only 22%. The reason for this is obviously the few non-zero components in the vector $\mathbf{y}_r$. Thus the QR-algorithm does not seem the most suitable procedure. Ward [25] has made a comparison with the procedure of Hyman. This will be examined in more detail in Section 4.2.6 and it will be seen that Hyman is preferable to QR when a vector (pipeline) computer is available.

4.2.5 The QR-algorithm for symmetrical tridiagonal matrices

A special form of the QR-algorithm is studied in this section. It is addressed specifically at problems containing a symmetrical tridiagonal matrix. Sameh and Kuck [23] have shown that a step in this version of the QR-procedure is equivalent to a linear recurrence for which it has shown in Chapter 3 that, when implemented on an SIMD computer, logarithmic performance is attainable.

As in the previous section we investigate the QR-algorithm with a shift of origin, the purpose of which is to obtain an improvement in the convergence rate. A double shift is not necessary since, in the case of symmetrical matrices $\mathbf{A}$, only real eigenvalues are relevant.

In the discussion that follows a QR step on the matrix $\mathbf{A}-t\mathbf{I}$ shown below is considered.

$$\mathbf{A}-t\mathbf{I} = \begin{bmatrix} a_1 & b_2 & & & \\ b_2 & a_2 & \ddots & & \\ & \ddots & \ddots & \ddots & \\ & & \ddots & & b_n \\ & & & b_n & a_n \end{bmatrix} \tag{4.100}$$

We seek a decomposition

$$\mathbf{Q}(\mathbf{A}-t\mathbf{I}) = \mathbf{R} \quad . \tag{4.101}$$

The orthogonal, unitary matrix $\mathbf{Q}$ can be chosen to be the product of $(n-1)$ rotation matrices. Thus

$$\mathbf{Q} = \mathbf{U}_{n-1,n}\,\mathbf{U}_{n-2,n-1}\,\ldots\,\mathbf{U}_{2,3}\,\mathbf{U}_{12} \quad . \tag{4.102}$$

As in the Jacobi procedure it will be supposed that $\mathbf{U}_{j,j+1}$ represents a rotation in position $(j, j+1)$. Then the left-multiplication

$$\mathbf{U}_{12}(\mathbf{A}-t\mathbf{I}) = \tag{4.103}$$

$$= \begin{bmatrix} c_1 & s_1 & & & \\ -s_1 & c_1 & & \mathbf{0} & \\ & & 1 & & \\ \mathbf{0} & & & \ddots & \\ & & & & 1 \end{bmatrix} \begin{bmatrix} a_1 & b_2 & & & \\ b_2 & a_2 & b_3 & & \\ & b_3 & a_3 & \ddots & \\ & & \ddots & \ddots & b_n \\ & & & b_n & a_n \end{bmatrix} = \begin{bmatrix} p_1 & q_2 & t_3 & & \\ 0 & g_2 & c_1 b_3 & & \\ & b_3 & a_3 & \ddots & \\ & & \ddots & \ddots & b_n \\ & & & b_n & a_n \end{bmatrix}$$

with $c_1/s_1 = \cos/\sin \phi_1$, is made to eliminate the element in position (2,1) by appropriate choice of the angle ϕ_1 (precisely $-a_1 \sin \phi_1 + b_2 \cos \phi_1 = 0$). Pre-multiplying by $\mathbf{U}_{23}$ we have

$$\mathbf{U}_{23}\mathbf{U}_{12}(\mathbf{A}-t\mathbf{I}) = \begin{bmatrix} 1 & 0 & 0 & & & \\ 0 & c_2 & s_2 & 0 & & \\ 0 & -s_2 & c_2 & & & \\ & & & 1 & & \\ & & & & \ddots & \\ 0 & & & & & 1 \end{bmatrix} \begin{bmatrix} p_1 & q_2 & t_3 & & & \\ 0 & g_2 & c_1 b_3 & 0 & & \\ & b_3 & a_3 & b_4 & & \\ & & & \ddots & \ddots & \\ & & & \ddots & & b_n \\ & & & & b_n & a_n \end{bmatrix}$$

$$= \begin{bmatrix} p_1 & q_2 & t_3 & & & \\ 0 & p_2 & q_3 & t_4 & & \\ 0 & 0 & g_3 & b_4 c_2 & & \\ & & b_4 & a_4 & b_5 & \\ & & & \ddots & \ddots & b_n \\ & & & & b_n & a_n \end{bmatrix}$$

where

$$p_1 = a_1c_1+b_2s_1 \qquad -a_1s_1+b_2c_1 = 0$$

$$q_2 = b_2c_1+a_2s_1 \qquad g_2 = -b_2s_1+a_2c_1$$

$$t_3 = s_1b_3 \;;$$

$$p_2 = g_2c_2+b_3s_2 \qquad -g_2s_2+b_3c_2 = 0$$

$$q_3 = c_2c_1b_3+a_3s_2 \qquad g_3 = -b_3c_1s_2+a_3c_2 \quad .$$

$$t_4 = b_4s_2 \quad .$$

We see that $\tan \Phi_1 = b_2/a_1$ so that $p_1 = \cos\phi_1\,(a_1+b_2^2/a_1) = (a_1^2+b_2^2)^{1/2}$ and $\sin\phi_1 = b_2/p_1, \cos\phi_1 = a_1/p_1$.

Similarly $\tan\phi_2 = b_3/g_2$ and $p_2 = (g_2^2+b_3^2)^{1/2}$ giving $\sin\phi_2 = b_3/p_2$, $\cos\phi_2 = g_2/p_2$.

We have shown two steps in the transformation chain (4.102). It is easily verified that at the rth step we have

$$p_r = g_rc_r+b_{r+1}s_r \qquad -g_rs_r+b_{r+1}c_r = 0$$

$$q_{r+1} = c_rc_{r-1}b_{r+1}+a_{r+1}s_r \qquad g_{r+1} = -b_{r+1}c_{r-1}s_r+a_{r+1}c_r \quad .$$

$$t_{r+2} = b_{r+2}s_r \quad . \tag{4.104}$$

With easily developable initial conditions these hold for $r = 1, 2, \ldots, n-1$. This was shown by Reinsch [30]. It follows easily by induction that at each stage the angle ϕ_r required to zeroise the element in position $(r+1, r)$ is given by

$$\tan\phi_r = b_{r+1}/g_r \quad . \tag{4.105}$$

Alternatively we can write

$$\sin\phi_r = \frac{b_{r+1}}{p_r}\,, \qquad \cos\phi_r = \frac{g_r}{p_r}\;.$$

Then we have

$$p_rc_r = g_r = a_rc_{r-1}-b_rs_{r-1}c_{r-2} = a_rc_{r-1}-c_{r-2}b_r^2/p_{r-1}\;. \tag{4.106}$$

Now let w_r be defined by

$$w_r := c_r\prod_{i=1}^{r} p_i \qquad (r = 1, 2, \ldots, n-1). \tag{4.107}$$

Then, using (4.106) we get for w_r a linear second order recurrence relation, namely

$$w_r = a_r w_{r-1} - b_r^2 w_{r-2} \qquad (r = 1, 2, \ldots, n-1) \qquad (4.108)$$

with $w_0=1$, $w_1=a_1$, $w_{-1}=0$.

(4.108) corresponds to the linear equation system

$$\mathbf{L}\,\mathbf{w} = \mathbf{i}_1 \quad ,$$

namely

$$\begin{bmatrix} 1 & & & & 0 \\ -a_1 & 1 & & & \\ b_2^2 & & \ddots & & \\ & \ddots & \ddots & & \\ & & b_{n-2}^2 & -a_{n-1} & 1 \end{bmatrix} \begin{bmatrix} w_0 \\ w_1 \\ \vdots \\ \\ w_{n-1} \end{bmatrix} = \begin{bmatrix} 1 \\ 0 \\ \vdots \\ \\ 0 \end{bmatrix} . \qquad (4.109)$$

In addition we have

$$1 = c_r^2 + s_r^2 = c_r^2 + \frac{b_{r+1}^2}{p_r^2}$$

so that, if

$$z_r^2 := \sum_{i=1}^{r} p_i^2 , \qquad (4.110)$$

we get

$$z_r^2 = c_r^2 z_r^2 + \frac{b_{r+1}^2}{p_r^2} z_r^2 = w_r^2 + b_{r+1}^2\, z_{r-1}^2 \qquad (r=1, 2, \ldots, n-1)$$

with $z_0^2 = 1$. (4.111)

This is a first order recurrence for z_r^2 and to it corresponds a system of linear equations

$$\begin{bmatrix} 1 & & & & 0 \\ -b_2^2 & 1 & & & \\ & & 1 & & \\ & & \ddots & \ddots & \\ & & & -b_n^2 & 1 \end{bmatrix} \begin{bmatrix} z_0^2 \\ z_1^2 \\ z_3^2 \\ \vdots \\ z_{n-1}^2 \end{bmatrix} = \begin{bmatrix} w_0^2 \\ w_1^2 \\ \\ \vdots \\ w_{n-1}^2 \end{bmatrix} . \qquad (4.112)$$

From the definitions we obtain the formulae

$$p_r^2 = z_r^2/z_{r-1}^2 \tag{4.113}$$

$$c_r^2 = w_r^2/z_r^2$$

$$s_r^2 = b_{r+1}^2/p_r^2, \qquad (r=1, 2, \ldots, n-1)$$

and the elements of the transformed matrix $\mathbf{A}' = \mathbf{Q}^T \mathbf{A} \mathbf{Q}$ are given by

$$a_r' = \frac{w_{r-1}\, w_r}{z_r^2}(b_{r+1}^2 + p_r^2) + a_{r+1}\, s_r^2 \tag{4.114}$$

$$b_{r+1}'^2 = p_{r+1}^2\, s_r^2 \qquad (r=1, 2, \ldots, n-1)$$

and

$$a_n' = \frac{w_{n-1}}{z_{n-1}^2}\, a_n - \frac{w_{n-2}\, w_{n-1}}{z_{n-1}^2}\, b_n^2 \quad .$$

As can be seen, during the whole evolution of the algorithm the neighbouring diagonal terms are needed only in squared form. This can be implemented (with $n-1$ processors) in a single step before the algorithm begins.

Sameh and Kuck [23] have shown that to calculate a', b'^2 from (4.114), five time steps are required with $2n$ processors. The main labour resides in the solution of the two linear recurrences (4.108) and (4.111). Discussions relevant to this are to be found in Chapter 3. Table 4.2 below gives the requirements for a QR step.

Table 4.2 Requirements of a QR step

Process	*Steps*	*Processors*
1. Solution of $\mathbf{L}\,\mathbf{w} = i_1$	$2 \log n$	$n-1$
2. $\hat{\mathbf{w}}^T = (w_0^2, \ldots, w_{n-1}^2)$	1	n
3. Solution of $\hat{\mathbf{L}}\,\hat{\mathbf{z}} = \hat{\mathbf{w}}$	2 log n	$2n-4$
4. Calculation of a', b'^2	5	$2n$
Total:	$6+4 \log n$	$2n$

$T_{2n} = 6 + 4 \log n.$

Reinsch [30] has shown that to implement the procedure sequentially requires $T_1 = 11n$ steps. Hence we obtain

Speed up $\qquad S_{2n} \doteq 11n/(4 \log n)$

Efficiency $\qquad E_{2n} \doteq 11/(8 \log n)$

This means, for example, with n=64, p=128, an efficiency of about 23%.

4.2.6 The Hyman procedure for upper Hessenberg matrices

The procedures discussed so far were all developed for an SIMD machine of the ILLIAC IV type. Ward [25] implemented the QR-algorithm on the CDC-STAR 100, a vector machine, and for comparison, the algorithm of Hyman, which is rarely employed on serial machines. According to Ward, Hyman's Algorithm is more efficient on a vector machine than the QR-algorithm. Ward did not parallelise the QR-algorithm to the extent ascribed to Kuck and Sameh [22], which has been described above in Section 4.2.4. However, the same essential disadvantage remains, namely the occurrence of relatively short vectors.

The following theory lies behind the Hyman algorithm. We suppose again that the basic matrix $\mathbf{A} = [a_{ij}\colon i, j=1, 2, \ldots n]$ is of upper Hessenberg form. Then the following system of equations is to be solved:

$$(\mathbf{A} - z\,\mathbf{I})\mathbf{y} = \mathbf{b} \tag{4.115}$$

being specifically of the form

$$\begin{bmatrix} * & & * & \\ * & * & & \\ & \ddots & \ddots & \\ & & * & * \end{bmatrix} \begin{bmatrix} * \\ * \\ \vdots \\ 1 \end{bmatrix} = \begin{bmatrix} f(z) \\ 0 \\ \vdots \\ 0 \end{bmatrix} ,$$

where z is a prescribed real number.

The last row gives

$$a_{n,\,n-1}y_{n-1} + (a_{nn}-z)\cdot 1 = 0 \tag{4.116}$$

and starting with this, the $y_i\,(i=n-1,n-2, \ldots, 1)$ can be calculated recursively until finally a value for $f(z)$ is obtained from the first equation. The general form for y_{n-i} is

$$a_{n-i+1,\,n-i}\,y_{n-i} = (z-a_{n-i+1,\,n-i+1})y_{n-i+1} - \sum_{j=n-i+2}^{n} a_{n-i+1,\,j}\,y_j \tag{4.117}$$

$$(i=1, 2, \ldots, n-1) \quad \text{with} \quad y_n=1 \;.$$

This is once again a linear recurrence for $y_1, y_2, \ldots, y_{n-1}$ and finally we have

$$(a_{11}-1)y_1+a_{12}y_2+a_{13}y_3+\ldots+a_{1n}y_n = f(z). \tag{4.118}$$

The special form of the set (4.115) makes it easy to identify the elements in the inverse of $\mathbf{A} - z\,\mathbf{I}$ and, in particular, solving for $y_n=1$, we get

$$1 = y_n = \frac{f(z)\,(-1)^{n-1} a_{21}a_{32} \ldots a_{n,n-1}}{\det(\mathbf{A} - z\,\mathbf{I})} \quad . \tag{4.119}$$

Alternatively, using the characteristic polynomial $\Phi(z) := \det(\mathbf{A} - z\,\mathbf{I})$ we have

$$f(z) = \frac{\Phi(z)\,(-1)^{n-1}}{a_{21}a_{32} \ldots a_{n,n-1}} \quad . \tag{4.120}$$

If it be supposed now that all the subdiagonal elements of $\mathbf{A}$ are non-zero (otherwise $\mathbf{A}$ is easily partitioned), the zeros of $f(z)$ are identical with those of the characteristic polynomial $\Phi(z)$.

Hence the eigenvalues can be calculated by finding the zeros of $f(z)$, for example by the usual Newton procedure, for the values of $f'(z)$ and higher derivatives are as easily calculated as $f(z)$.

Parallelisation on a vector processor depends again very heavily on the organisation of the data in store. Indeed it is here only a question as to whether it is more suitable to store the matrix elements by rows or by columns. On the CDC-STAR, in particular, vector operations can be carried out only on data stored sequentially. The implementation of multiplying a matrix $\mathbf{M}$ with a vector $\mathbf{x}$ depends therefore on the manner in which the elements of $\mathbf{M}$ are stored. Thus, to get $\mathbf{y} = \mathbf{M}\,\mathbf{x}$:

(a) in the case of storage by rows n inner products are formed to give

$$y_i = \sum_{j=1}^{n} m_{ij}\,x_j \qquad (i = 1, 2, \ldots n).$$

If the inner product operation be denoted by $\circledast$ this could be written:

```
for    i=1  to n do
           y [i]  :=  M[i,] ⊛ x
end
```

where $M[i,]$ denotes the ith row vector.

(b) in the case of storage by columns the sums of products $\sum_{i=1}^{n} m_{ij}\,x_i$ are formed and added for $j=1, 2, \ldots, n$, giving finally the whole vector y.

This entails n products of elements of the column vectors with a scalar (element of **x**) and n vector additions. Using a similar, self-explanatory, computer-like notation the procedure can be represented by

```
        y := vector (0)
for     j = 1 to n do
        y := y + x [j] * M[,j]
end
```

Here $M[,j]$ denotes the column vectors of **M**.

Let us calculate the time requirements of these operations. For a pipeline computer the time requirements of vector operations need to be enumerated explicitly since they depend on the length of the vector under consideration. Generally, the time T_{op} required for a vector operation op is given by

$$T_{op} = \sigma_{op} + \tau_{op}\, n$$

where σ_{op} = start up time; τ_{op} = incremental time; n = length of the vector.

For the CDC-STAR 100 an idea of the time required by simple operations is as shown in Table 4.3. Note that the time unit is a 'cycle', that is 40×10^{-9} seconds. It can be seen that for small n the duration of the double starting time σ_{op} in (b) is predominant and thus row vector storage in this case is preferable.

Table 4.3

	Scalar op	σ_{op}	τ_{op}	Symbol
Addition/subtraction	13	71	0.5	+ −
Multiplication	17	159	1	*
Division	46	167	2	/
Sum	–	122	4	Σ
Inner product	–	137	5.5	⊛

For large n, however, the comparison

$$\tau_{\circledast} > \tau_{+} + \tau_{*}$$

is the significant factor. For the implementation of the Hyman procedure multiplications of ⊛ type predominate. This means that, for large n, columnar storage of the matrix elements is preferable.

According to Ward [25], a Hyman step entails $0(n)$ vector operations as against $0(n^2)$ for a single QR step. We note that for each eigenvalue fewer QR than Hyman steps are needed. According to the experiments of Ward the requirement is roughly.

$1\frac{1}{3}$ QR steps versus 7 Hyman steps.

If this is the case it follows that for matrices whose dimensions exceed 27×27 the Hyman procedure is preferable.

4.2.7 Simultaneous iteration for the calculation of the r largest eigenvalues

In many cases the problem is not to calculate all the eigenvalues but rather the largest in absolute value. A simultaneous iteration procedure allows, for arbitrary r $(1 \leqslant r \leqslant n)$, the r largest eigenvalues (in absolute value) of a symmetrical positive definite, $n \times n$ matrix to be calculated.

Theorem (4.121)
Let $\mathbf{A}$ be a symmetric, positive-definite $n \times n$ matrix with eigenvalues

$$|\lambda_1| > |\lambda_2| > \ldots > |\lambda_r| > |\lambda_{r+1}| \geqslant |\lambda_{r+2}| \geqslant \ldots \geqslant |\lambda_n| .$$

Let $\mathbf{X}_0$ be an orthonormal matrix ($\mathbf{X}_0^T \mathbf{X}_0 = \mathbf{I}$) of dimension $n \times r$ used to start the procedure. It has rank r.

For $k = 1, 2, \ldots$ let

(i) $\mathbf{Z}_k := \mathbf{A}\,\mathbf{X}_{k-1}$

(ii) $\mathbf{Z}_k = \mathbf{X}_k\,\mathbf{R}_k$ be a decomposition of $\mathbf{Z}_k$ into an upper $r \times r$ triangular matrix $\mathbf{R}_k$ and an orthonormal $n \times r$ matrix $\mathbf{X}_k$, thus

(iii) $\mathbf{X}_k := \mathbf{Z}_k\,\mathbf{R}_k^{-1}$.

Then

$$\lim_{k\to\infty} \mathbf{X}_k = [\mathbf{z}_1, \mathbf{z}_2, \ldots, \mathbf{z}_r] \; ;$$

$$\lim_{k\to\infty} \mathbf{R}_k = \begin{bmatrix} \lambda_1 & & \mathbf{0} \\ & \lambda_2 & \\ & & \\ & \mathbf{0} & & \lambda_n \end{bmatrix}$$

where $\mathbf{z}_i$ is the eigenvector corresponding to the eigenvalue λ_i $(i=1, 2, \ldots, n)$. For the *proof* see Blumenfeld [31].

The decomposition in the second step can be implemented by the Gram–Schmidt procedure.

Let

$$\mathbf{Z}_k = \mathbf{Z} = [z_1, \ldots, z_r] \; , \qquad \mathbf{R} = [r_{ij}]$$

$$\mathbf{X}_k = \mathbf{X} = [x_1, \ldots, x_r] \; .$$

x_i and r_{ij} are then to be calculated from z_i as follows:

(i) For l=1,

$$x_1 := z_1/r_{11}$$
$$r_{11} := (z_1^{\mathrm{T}} z_1)^{1/2}$$

(ii) For l=2, . . . , r

$$x_l := (z_l - \sum_{i=1}^{l-1} r_{il} x_i)/r_{ll}$$

$$r_{il} := x_i^{\mathrm{T}} z_l \quad (i=1, 2, \ldots, l-1).$$

$$r_{ll} := \left\{ (z_1 - \sum_{i=1}^{l-1} r_{il} x_i)^{\mathrm{T}} (z_l - \sum_{i=1}^{l-1} r_{il} x_i) \right\}^{1/2} .$$

The algorithm terminates when the sum of the absolute values of the non-diagonal elements of $\mathbf{R}_k$ falls below a prescribed limit.

Schendel, Gomm and Weistroffer [26] have implemented this algorithm for the general model of a vector computer taking into account vector operations on vectors with zero components. The numerical results encourage belief in the stability of the algorithm.

This algorithm is already in a form suitable for parallel implementation since it is given in a vector formulation. To carry out the implementation a computer must be able to perform the following operations in parallel:

(a) componentwise addition, subtraction, multiplication and division of two vectors;
(b) summation of all components of a vector;
(c) application of a scalar to all the components of a vector.

In addition, for optimal parallel implementation.

(d) componentwise computation of the square root of a vector.

Remark

The formation of the scalar product of two vectors consists of (a) and (b) in succession. (c) makes possible operations between scalars and vectors. Because of (b), and the fact that the dimension of the vector operand changes at each step of the orthonormalisation procedure, programming on a vector (pipeline) computer is feasible. A program written in Fortran for the CDC-STAR is given in [32].

4.3 Nonlinear Problems

We intend in this section to examine the solution of nonlinear equations. The solutions are to be obtained by using suitable parallel algorithms on a parallel machine. In this connection there will be some discussion of search methods.

The basic problem is that of calculating a zero of a real function f. A starting interval $I^{(0)}$ known to contain the zero z is chosen and algorithms are sought to generate a sequence of nested intervals $\{I^{(k)}\}$ such that for each $I^{(k)}$, $z \in I^{(k)}$. The aim will be to obtain monotonic algorithms. At each iteration step n function values will be calculated in parallel using n processors.

More formally we are given a real function

$$f : [x_1^{(0)}, x_2^{(0)}] \rightarrow \mathbb{R} \quad ,$$

which possesses a zero z in the interval $I^{(0)} := [x_1^{(0)}, x_2^{(0)}]$. The search for solutions of the equation $f(x) = 0$ can in this case be expressed as:

Generate a sequence of iterated intervals $\{I^{(k)}\}$ where $I^{(k)}$ is an improvement on the initial interval in the following sense:

$$z \in I^{(k)} \quad , \qquad k \geqslant 0 \tag{4.121}$$

$$I^{(0)} \supset I^{(1)} \supset I^{(2)} \ldots \tag{4.122}$$

$$I^{(k)} \rightarrow z \text{ for } k \rightarrow \infty \quad . \tag{4.123}$$

These three equations entail that the algorithm generates a monotonic sequence of lower and upper bounds for z. Moreover convergence must be assured independently of how well the initial interval $I^{(0)}$ is chosen. In practice the requirement (4.123) is introduced into the algorithm in the following weaker form:

$$I^{(k)} = I, \qquad k \geqslant k_0 \quad ,$$

because, among other things, of accumulated rounding errors. The interval is a kind of numerical fixed point which cannot be improved upon by the method considered. In order to be able to measure the sharpness of the bounds given by the interval $I = [x_1, x_2]$ we consider the interval length $d(I) := x_2 - x_1$. Convergence conditions for the sequence $\{I^{(k)}\}$ are established through the corresponding conditions for the sequence $\{d(I^{(k)})\}$.

4.3.1 Bisection procedure

To determine a zero of the function f: $[a, b] \rightarrow \mathbb{R}$ by the bisection procedure (interval reduction) a series of intervals $I_k = [a_k, b_k]$ $(a_k, b_k \in \mathbb{R})$ each containing the zero is generated by continued halving in such a way that

$$f(a_k) f(b_k) < 0 \quad . \tag{4.124}$$

Parallelisation

Let p be the number of processors. For parallel implementation the intervals are divided into $p+1$ subintervals. There are then p interior points the function values at which can be calculated in parallel. Out of the $p+1$ intervals a particular interval $I_k = [a_k, b_k]$ is sought for which (4.124) holds. If no such interval exists, one of the points of division must be a zero. It will be assumed that the search time for the new subinterval can be neglected by comparison with the time required to calculate the function values. This assumption seems justified in view of the likelihood that such a slowly convergent procedure as interval reduction would be contemplated only when the function is a complicated one. Under this assumption we estimate for the speed up ratio:

$$S_p = k_1/k_p \ , \tag{4.125}$$

where k_i is the number of interval subdivisions required when i processors are used.

If, using the algorithm serially, we want to find z^* such that $|z-z^*| < \epsilon$, where z is the true value of the zero, then the final interval must be smaller than 2ϵ in length. Let d be the length of the starting interval $I_0 = [a, b]$. Then

$$d/2^{k_1} \leqslant 2\epsilon \ .$$

Since the parallel version requires a $(p+1)$-fold reduction in the interval length we have

$$d/(p+1)^{k_p} \leqslant 2\epsilon$$

and thus

$$2^{k_1} \doteqdot (p+1)^{k_p}$$

or

$$k_1 \doteqdot k_p \log_2 (p+1) \ .$$

It follows that

$$S_p \doteqdot \log_2 (p+1) \ ,$$

that is

$$S_p = 0 (\log_2 p) \ . \tag{4.126}$$

If the termination criterion is $|f(z^*)| < \epsilon$, then again (4.126) is obtained (see Shedler [33]).

4.3.2 Regula falsi

Regula falsi also entails the generation of a sequence $I_k = [a_k, b_k]$ of intervals containing the zero in question, so that (4.124) must hold. The procedure is effectively to pass a straight line through the points $(a_k, f(a_k)), (b_k, f(b_k))$, and to use the intersection of this line with the x-axis as the upper or lower boundary of the next interval according to (4.124).

Parallel version
I_k is again subdivided by p equidistant points into $p+1$ equal subintervals and the function values $f(m_i)$ at the intermediate points m_i $(i=1, 2, \ldots, p)$ are calculated simultaneously in parallel. The regula falsi procedure is then applied for $i=1, 2, \ldots, p$ to the subintervals (a_k, m_i) or (m_i, b_k) accordingly as $f(a_k)\, f(m_i) < 0$ or $f(m_i)\, f(b_k) < 0$, and if neither is true then of course m_i is the zero in question. This gives a sequence of p points z_i $(i=1, 2, \ldots, p)$. Finally, out of a_k, b_k, m_i, z_i $(i=1, 2, \ldots, p)$ that pair is chosen which defines the shortest subinterval and contains the zero. This defines $[a_{k+1}, b_{k+1}]$. Figure 4.3 illustrates this for $p=4$. Obviously the choice $I_{k+1} = [m_1, z_2]$ is made.

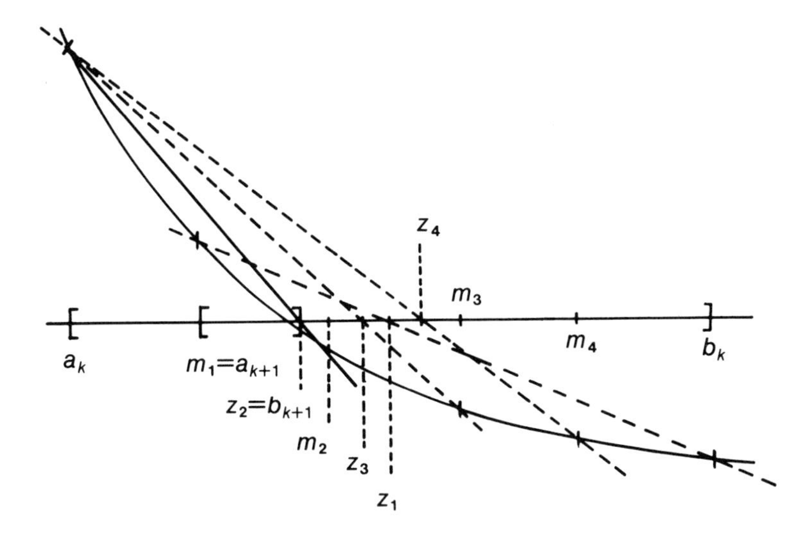

Fig. 4.3 – Regula falsi for $p=4$.

General statements about speed-up cannot be made for this procedure. For a sensible speed-up it is also the case here that the time required to calculate the function values must be considerably greater than the search time for the next interval. Empirical results are to be found in Shedler [33].

4.3.3 An iterative parallel procedure for the location of zeros

Some definitions, ideas and results from interval arithmetic are required for the construction of the next algorithm.

Let X and Y be two finite intervals of the real axis. The following operations for X and Y are defined.

$$X \otimes Y := \{x \otimes y \mid x \in X, y \in Y\}, \tag{4.127}$$

where $\otimes$ is one of the operators $\{+, -, *, /\}$, and 0 is excluded from Y when the operation is a division.

Let

$$\mathscr{I}(\mathbb{R}) := \{[a, b] \mid a, b \in \mathbb{R}\}$$

and $X, Y \in \mathscr{I}(\mathbb{R}), X = [a, b], Y = [c, d]$. Then (4.127) gives specifically

$$X+Y = [a+c, b+d]$$
$$X-Y = [a-d, b-c]$$
$$X*Y = [\min(ac, ad, bc, bd), \max(ac, ad, bc, bd)]$$
$$X/Y = [a, b]\cdot[1/d, 1/c], \qquad (0 \notin Y).$$

Interval arithmetic does not always share the properties of familiar arithmetical operations. It follows that in sequences of operations involving parentheses the order must be uniquely prescribed. The set of intervals with a single point $\{[a, a] \mid a \in \mathbb{R}\}$ is under interval arithmetic isomorphic with $\mathbb{R}$. $[a, a]$ will be identified with a.

The following Theorem is fundamental.

Theorem (4.128)
Let $F(X_1, X_2, \ldots, X_n)$ be a rational expression in the interval variables $X_1, X_2, \ldots, X_n$, that is to say a finite combination of the variables $X_1, X_2, \ldots, X_n$ and a finite set of constant intervals with interval arithmetical operations. It follows from

$$X_i' \subseteq X_1, \ldots, X_n' \subseteq X_n$$

that

$$F(X_1', \ldots, X_n') \subseteq F(X_1, \ldots, X_n)$$

for each set of intervals $X_1, \ldots, X_n$, for which the interval arithmetic operations on F are defined. If, in particular, $X_1', \ldots, X_n'$ and the constant intervals in F are real numbers, so also is the value $F(X_1', \ldots, X_n')$ a real number contained in the interval $F(X_1, \ldots, X_n)$. For a proof see Moore [34].

Interval arithmetic is an extension of arithmetic operations on the elements of intervals. A corresponding extension will now be given for real continuous functions. In this way we obtain an *interval analysis* in addition to interval arithmetic. For this purpose let again $\mathscr{I} = \mathscr{I}(\mathbb{R})$ be the set of all real closed intervals, $\mathscr{I}_I$ the set of all real, closed subintervals of a given interval I, which means

$$\mathscr{I}_I = \{J \mid J \in \mathscr{I}, J \subseteq I\} .$$

Definition (4.129)
Let $f : I \to \mathscr{J}$ be a continuous interval-valued function on a closed real interval I.† Then there exists a natural extension of f to an interval-valued function J_I. We define the extension $\bar{f}$ of f by

$$\bar{f} : \mathscr{J}_I \to \mathscr{J} ,$$

where

$$\bar{f}(X) = \bigcup_{x \in X} f(x) \qquad (x \in \mathscr{J}_I) \ .$$

Thus $\bar{f}(X)$ is the union of values $f(x)$ for all x in the subinterval X of the interval I.

Note: The operations of interval arithmetic are merely the extension of real arithmetic operations to take account of all points in the intervals.

It will now be shown how bounds can be calculated for the intervals $\bar{f}(x)$ by using interval arithmetic. In other words we shall show how to compute the maximum and minimum of the interval f on X. Let $f(x) : I \to \mathbb{R}$ be a finite real, rational function.‡ Replace x in $f(x)$ formally by an interval X. Then an interval valued function $F(X)$ is obtained,

$$F(X) : \mathscr{J}_I \to \mathscr{J} .$$

By Theorem (4.128) we have

$$F(X) \supseteq \bar{f}(X) \ . \tag{4.130}$$

Now let $X = [a, b]$ be subdivided as follows.

$$[a, b] = [a, a_1] \cup [a_1, a_2] \cup \ldots \cup [a_n, b] .$$

Then, by (4.130),

$$F([a_i, a_{i+1}]) \supseteq \bar{f}([a_i, a_{i+1}])$$

$$(i=0, \ldots, n; \ a_0=a, a_{n+1}=b).$$

Thus

$$\bigcup_{i=0}^{n} F([a_i, a_{i+1}]) \supseteq \bar{f}([a, b]) \ .$$

† If the intervals are all of form $[a, a] = a$, then f can be identified with a real-valued function.
‡ This is, in practice, no restriction since any real function is approximated by a finite rational function.

It follows that $\bigcup_{i=0}^{n} F([a_i, a_{i+1}])$ is in general a better estimate of $\bar{f}([a,b])$ than $F([a,b])$.

Example
Let $f(x) = x^2-2x$ and the interval $X=[-2,2]$. The least value of $f(x)$ in X is $f(+1) = -1$. The maximum is $f(-2)=8$. Thus $\bar{f}(X) = [-1,8]$, and

$$F(X) = F([-2,2)] = X \cdot X - 2 \cdot X = [-2,2]*[-2,2]-2*[-2,2]$$
$$= [-4,4]-[-4,4] = [-8,8] .$$

Using the decomposition $X = [-2,0] \cup [0,2]$ we get

$$F([-2,0]) = [-2,0]*[-2,0]-2[-2,0]=[0,4]-[-4,0]=[0,8] .$$
$$F([0,2]) = [0,2]*[0,2]-2[0,2]=[0,4]-[0,4]=[-4,4]$$
$$F([-2,0]) \cup F([0,2]) = [0,8] \cup [-4,4] = [-4,8] \subset F([-2,2]) .$$

The following theorem gives a convergence criterion for the assertion made before the example.

Theorem (4.131)
Let $f(x)$ be a real, rational function and $F(X)$ the corresponding interval function. Let $\bar{f}(X)$ be the extension of $f(x)$, and let X be expressed as the union

$$X = \bigcup_{i=1}^{n} X_i$$

where the length $d(X_i)$ of the interval X_i is given by

$$d(X_i) = d(X)/n .$$

Then there exists a positive real number c with the property that

$$\bigcup_i F(X_i) = \bar{f}(X) + E_n$$

where E_n is an interval having the property

$$0 \in E_n \quad \text{and} \quad d(E_n) \leqslant \frac{c}{n} d(X) .$$

For proof see Moore [34].

Interval version of the Newton procedure
An interval-arithmetic oriented version of the Newton procedure for the determination of a zero of a real rational function $f(x)$ is given by

$$N(X_i) = m(X_i) - \bar{f}(X_i)/\bar{f}'(X_i) \quad ,$$
$$X_{i+1} = N(X_i) \cap X_i, \qquad (i=0, 1, \ldots). \tag{4.132}$$

Here X_0 is the initial interval of the iteration and $m(X_i)$ is a point in X_i (for example the midpoint). Since $\bar{f}(X_i)$ and $\bar{f}'(X_i)$ are not in general known, they can be estimated by the interval functions $F(X_i)$ and $F'(X_i)$, where F is the interval function that corresponds to f. Then, if X is subdivided into subintervals, Theorem (4.131) can be used to obtain an improved estimate. The calculation of all $F(X_i)$ can be carried out simultaneously by parallel operation.

Description of the algorithm
The algorithm to be considered now is due to Herzberger [35]. We start with an interval $I^{(0)} = [x_1^{(0)}, x_2^{(0)}]$ as giving bounds for x. The iteration consists essentially of the following three steps:

Step 1. Choose k different points

$$x^{(0,1)}, x^{(0,2)}, \ldots, x^{(0,k)} \in I^{(0)} \quad .$$

Step 2. Calculate $f(x^{(0,i)})$ for $i=1, 2, \ldots, k$, which can be done in parallel.
Step 3. Compute an improved interval

$$I^{(1)} \subset I^{(0)}$$

such that

$$z \in I^{(1)} \quad .$$

Steps 1, 2 and 3 are repeated until $d(I^{(n)}) < h$, a prescribed quantity. The most exacting step is 3 which must ensure monotonicity.

Let $f \in C^{(k)}$ and let H, K be interval bounds for $\bar{f}'(I^{(0)})$ and $\overline{f^{(k)}}(I^{(0)})$, namely

$$H \supseteq f'(I^{(0)}), \qquad 0 \notin H,$$
$$K \supseteq \overline{f^{(k)}}(I^{(0)}).$$

Estimates for H and K can be obtained again from

$$H = F'(I^{(0)})$$
$$K = F^{(k)}(I^{(0)}),$$

supposing always that $0 \notin F'(I^{(0)})$. No further restrictions are imposed on H and K.

By using the function values computed at Step 2 it is possible to construct an interpolation polynomial $P_{k-1}(x)$ on the set of points $(x^{(0,i)}, f(x^{(0,i)}))$, $(i=1, 2, \ldots, k)$. $z^{(0)}$ denotes a real zero of $P_{k-1}(x)$ in $I^{(0)}$.

For this purpose we choose

$$x^{(0,1)} := x_1^{(0)}, \qquad x^{(0,k)} := x_2^{(0)} \quad .$$

In this way a change of sign of the interpolation polynomial will be forced. The remainder term (for polynomial interpolation) is given by

$$f(x) - P_{k-1}(x) = \frac{f^{(k)}(\zeta)}{k!} \prod_{i=1}^{k} (x - x^{(0,i)})$$

for $\zeta \in [x^{(0,1)}, x^{(0,k)}]$.

If the interval K be substituted for $f^{(k)}(\zeta)$ then we get for $x = z^{(0)}$ the interval expression

$$F^{(0)} = \frac{K}{k!} \prod_{i=1}^{k} (z^{(0)} - x^{(0,i)}). \tag{4.133}$$

By Theorem (4.128) $F^{(0)}$ contains the value $f(z^{(0)})$. Thus $z^{(0)}$ and $F^{(0)}$ can be used to initiate an interval-oriented Newton step, giving

$$X^{(1)} := (z^{(0)} - F^{(0)}/H) \cap X^{(0)}$$

where $z \in X^{(1)} \subset X^{(0)}$.

The following theorem can now be enunciated.

Theorem (4.134)
With the foregoing assumptions the iteration procedure using steps 1, 2, 3 generates a sequence of intervals $\{X^{(i)}\}$ with properties (4.121) to (4.123). The order of the convergence is at least equal to the number k of parallel function evaluations.

Sketch of proof Properties (4.121) to (4.123) can be inferred from (4.133) (see the literature). For the order of convergence we write:

$$\begin{aligned} \mathrm{d}(X^{(n+1)} &\leqslant \mathrm{d}(z^{(n)} - F^{(n)}/H) = \mathrm{d}(F^{(n)}/H) \\ &= \mathrm{d}\, \frac{K}{k!} \left(\prod_{i=1}^{k} (z^{(n)} - x^{(n,i)})/H \right) \\ &= c \prod_{i=1}^{k} (z^{(n)} - x^{(n,i)}) \leqslant c\, d\, (X^{(n)})^k \quad . \end{aligned}$$

4.3.4 Search procedures for the determination of the zeros of special classes of functions

Let $W_n = W_n[0,1]$ be the space of functions defined on $[0,1]$, whose $(n-1)$th derivatives are absolutely continuous, and whose nth derivatives are bounded almost everywhere. We seek a zero in $[0,1]$ of a function $f \in W_n$, for which

$$m \leqslant f^{(n)}(x) \leqslant M$$

for all $x \in [0,1]$, $m, M \in \mathbb{R}$.

Suppose that N tabular points $x_1, x_2, \ldots, x_N$ are given in $[0,1]$ at which the function values $y_i = f(x_i)$ $(i=1, 2, \ldots, N)$ are known. A class of functions containing f will now be defined.

Definition (4.135)
F is the class of functions defined as follows.

$$F = F_{n,N}(m,M) := \{f \in W_n \mid m \leqslant f^{(n)}(x) \leqslant M \quad \text{for all } x \in [0,1], f(x_i) = y_i, i=1, 2, \ldots, N\}.$$

We shall now introduce two extreme elements $\bar{s}(x)$ and $\underline{s}(x)$ with respect to F.

Definition (4.136)

$$\bar{s}(x) = \sup_{f \in F} f(x), \quad \underline{s}(x) = \inf_{f \in F} f(x).$$

$\bar{s}(x)$ and $\underline{s}(x)$ can be referred to respectively as the upper and lower *envelopes* or *hulls* of F.

Naturally for each $f \in F$ and $x \in [0,1]$ we have

$$\underline{s}(x) \leqslant f(x) \leqslant \bar{s}(x) \tag{4.137}$$

and for the zeros $\bar{z}, z, \underline{z}$ of $\bar{s}, f, \underline{s}$ respectively,

$$\bar{z} \leqslant z \leqslant \underline{z} \tag{4.138}$$

Thus the zeros of $\bar{s}$ and $\underline{s}$ are bounds for the zero z which is being sought. To understand the algorithm which follows, the following Theorem is important (see, for example, de Boor [36]).

Theorem (4.139)
Let G be the class of functions defined as follows:

$$G \equiv G_{n,N}(\mu) = \{g \in W_n \mid \; |g(x)| \leqslant \mu \text{ for all } x \in [0,1], \quad g(x_i) = y_i, i=1, 2, \ldots, N\}.$$

If G is not empty there exist two functions $g_1, g_2 \in G$ such that, for all $x \in [0,1]$ and all $g \in G$,

$$g_1(x) \leqslant g(x) \leqslant g_2(x).$$

g_1 and g_2 are *perfect splines* (see [36]†) of degree n with $k := N-n$ mesh points. This means that both are splines of order $n+1$ with k simple inner mesh points and a constant nth derivative

$$|g_1^{(n)}(x)| = |g_2^{(n)}(x)| = \mu$$

for all $x \in [0,1]$.

By analogy we now introduce perfect (m,M)-splines (Micchelli and Miranker [37]).

Definition (4.140)
Let mesh points $\xi_i \in [0,1]$, $i=0, 1, \ldots, k+1$ $(k := N-n)$ with $\xi_0 = 0$, $\xi_{k+1} = 1$ be given. Let $S_m(x) \in F$ and suppose that the following holds:

$$S_m^{(n)}(x) = \begin{cases} m, \; x \in [\xi_{k-i}, \xi_{k+1-i}], & \textit{even } i \\ M & \text{otherwise.} \end{cases}$$

Let $S_M(x) \in F$ be defined analogously, m and M being merely exchanged. Then the mesh points ξ_i^* of S_M can be different from the ξ_i pertaining to S_m. $S_m(x)$ and $S_M(x)$ are called *perfect* (M,m)*-splines.* When $m = -M$ the definition reduces to the usual definition for perfect splines.

In order to make use of (4.138) it is necessary first to determine $\underline{s}$ and $\bar{s}$. The following Theorem establishes the relationship between $\underline{s}$, $\bar{s}$, S_m and S_M.

Theorem (4.141)
Under the assumption that $F \neq \phi$ and that $S_m(x)$ and $S_M(x)$ are continuous in m and M, we have

$$\underline{s}(x) = \min(S_m(x), S_M(x)) \;;$$

$$\bar{s}(x) = \max(S_m(x), S_M(x)) \;.$$

Using this as a basis we obtain the following algorithm.

(a) Divide an interval $[a,b]$ known to contain the zero by N tabular points $x_1, x_2, \ldots, x_N$.
(b) Determine the mesh points $\xi_0, \xi_1, \ldots, \xi_{k+1}$ and $\xi_0^*, \xi_1^*, \ldots, \xi_{k+1}^*$ and the splines S_m and S_M.
(c) Determine $\underline{s}$ and $\bar{s}$.

† The basic text, is: *The Theory of Splines and their Applications,* Ahlberg, J. H., Nilson, E. N., and Walsh, J. L. New York: Academic Press, 1967.

(d) Calculate the zeros $\bar{z}$ and $\underline{z}$ of $\bar{s}$ and $\underline{s}$. Then $[\underline{z}, \bar{z}]$ is the new starting interval so return to (a).

To illustrate the procedure we give a simple and practically important example.

The parallelogram procedure

Taken N=2 and n=1. Thus a function $f \in W_1\ [0,1]$ may be thought of as given, whose zero in $[0,1]$ is sought, and which satisfies

$$m \leqslant f'(x) \leqslant M \qquad (\text{all } x \in [0,1]) \ .$$

Also it is supposed that N=2 points (x_1,y_1), (x_2,y_2) are given, where $y_1 = f(x_1)$, $y_2 = f(x_2)$, and that the zero z lies in $[x_1, x_2]$.

By a simple adaptation of (4.140), $S_m(x)$ has the property

$$\begin{aligned} S_m'(x) &= m \text{ for } x \in [\xi_1, \xi_2] = [\xi_1, x_2], \\ S_m'(x) &= M \text{ for } x \in [\xi_0, \xi_1] = [x_1, \xi_1]. \end{aligned} \tag{4.142}$$

Hence

$$\begin{aligned} S_m(x) &= mx + c_1 \quad \text{for} \quad x \in [\xi_1, x_2] \\ S_m(x) &= Mx + c_2 \quad \text{for} \quad x \in [x_1, \xi_1], \end{aligned}$$

c_1 and c_2 being real constants.

By continuity at ξ_1,

$$m\,\xi_1 + c_1 = M\xi_1 + c_2, \tag{4.143}$$

and because $S_m \in F$,

$$S_m(x_1) = y_1, \qquad S_m(x_2) = y_2. \tag{4.144}$$

These determine c_1 and c_2. Thus

$$\begin{aligned} S_m(x) &= M(x-x_1)+y_1 \quad \text{for } x \in [x_1, \xi_1] \\ S_m(x) &= m(x-x_2)+y_2 \quad \text{for } x \in [\xi_1, x_2], \end{aligned} \tag{4.145}$$

while (4.143) gives for ξ_1

$$\xi_1 = (y_2-y_1+Mx_1-mx_2)/(M-m) \ .$$

By analogy we have

$$\begin{aligned} S_M(x) &= m(x-x_1)+y_1 \quad \text{for } x \in [x_1, \xi_1^*] \\ S_M(x) &= M(x-x_2)+y_2 \quad \text{for } x \in [\xi_1^*, x_2] \end{aligned} \tag{4.146}$$

$$\xi_1^* = (y_2-y_1 + mx_1-Mx_2)/(m-M).$$

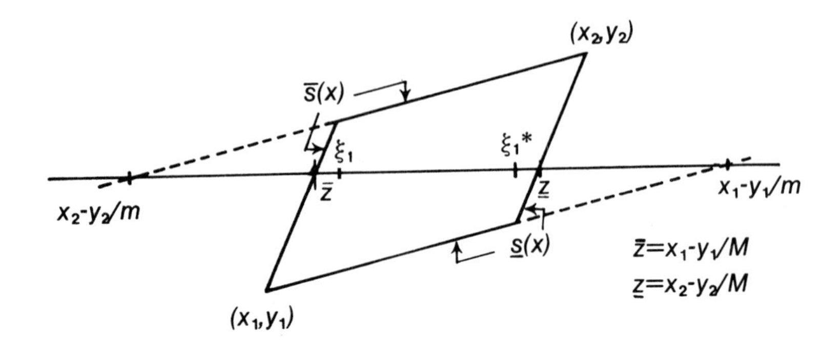

Fig. 4.4 Illustrating a step of the parallelogram procedure.

Figure 4.4 illustrates $s_m(x)$ and $s_M(x)$. It is obvious that $\underline{s}(x) = S_M(x)$ and $\bar{s}(x) = S_m(x)$. As shown, both m and M are positive and so the zeros $\underline{z}$ and $\bar{z}$ of $\underline{s}(x)$ and $\bar{s}(x)$ in $[x_1,x_2]$ are those shown, namely $\bar{z} = x_1 - y_1/M$, $\underline{z} = x_2 - y_2/M$, and since $\bar{z} < \underline{z}$ the interval for the next iteration is $(\bar{z}, \underline{z})$.

Convergence

Referring again to Fig. 4.4 we see that

$$\theta := \frac{\underline{z}-\bar{z}}{x_2-x_1} = \frac{x_2-x_1-\frac{1}{M}(y_2-y_1)}{x_2-x_1} = 1-\frac{1}{M}\frac{y_2-y_1}{x_2-x_1} ,$$

and since under the conditions shown it is clear that $(y_2-y_1)/(x_2-x_1) > m$ we have $\theta < 1-m/M$ indicating geometric convergence. This will always be the case as long as m and M have the same sign. The bisection procedure endeavours by twofold evaluation of the function to reduce the interval by a factor of $\theta_b = 1/4$. To improve on this using the parallelogram procedure we must have $m/M > 3/4$.

When z is a simple zero of f and $f \in C^2\ [0,1]$ the algorithm can be made to yield quadratic convergence by 'updating' m and M at each step. Take as an example

$$m := \min_{x \in I} f'(x), \qquad M := \max_{x \in I} f'(x),$$

where $I := [x_1,x_2]$. In this case the Mean Value Theorem shows that

$$\underline{z} - \bar{z} \leqslant (x_2-x_1)^2 \frac{\max_{\xi \in I} f''(\xi)}{\max_{\xi \in I} f'(\xi)} .$$

The parallel parallelogram procedure

In the case just considered (N=2, k=n=1) the degree of parallelism is rather slight, but if we increase the number N of tabular points, by analogy with (4.145) we get for the mesh points ξ_i of $S_m(x)$

$$\xi_i = (m\,x_{i+1} - Mx_i + y_i - y_{i+1})/(m-M), \qquad (i=k, k+2, \ldots)$$

$$\xi_i = (m\,x_i - Mx_{i+1} + y_{i+1} - y_i)/(m-M), \qquad (i=k-1, k-3, \ldots).$$

The mesh points ξ_i^* of $S_M(x)$ are again obtained by interchanging m and M. $S_m(x)$ and $S_M(x)$ are linear functions on the intervals $[\xi_i, \xi_{i+1}]$ (i=0, 1, . . .N–1) which interpolate the function values between the mesh points and have slopes m and M.

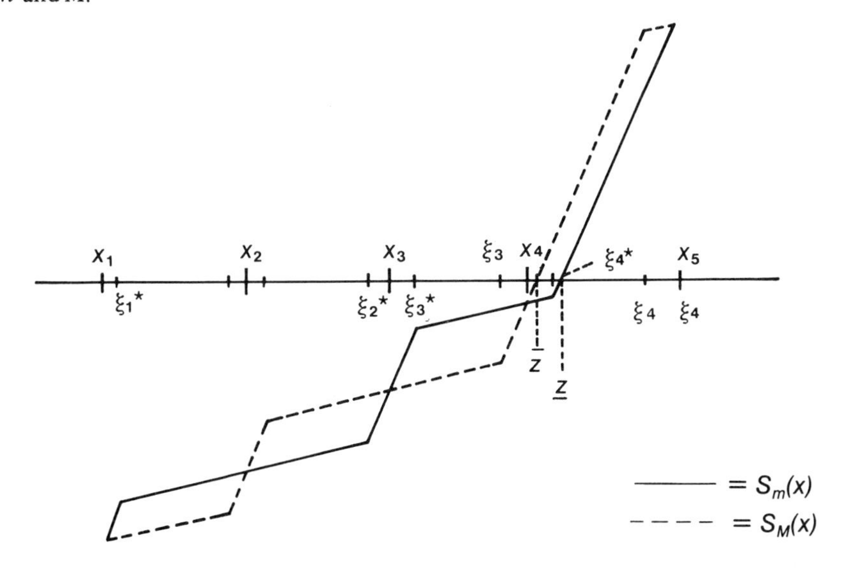

Fig. 4.5 – Illustrating parallel parallelogram procedure.

Figure 4.5 illustrates that

$$\underline{s}(x) = \min(S_m(x), S_M(x)) = \begin{cases} S_m(x), x \in [x_{N-i}, x_{N-i+1}] & (i \text{ even}); \\ S_M(x) & \text{otherwise}; \end{cases}$$

$$\bar{s}(x) = \max(S_m(x), S_M(x)) = \begin{cases} S_m(x), x \in [x_{N-i}, x_{N-i+1}] & (i \text{ odd}); \\ S_m(x) & \text{otherwise}. \end{cases}$$

Optimal choice of mesh

We showed how to determine the mesh ξ_i (i=1, . . . , k) of the splines. However, the choice of the x_i (i=1, 2, . . . , N) mesh was left open. Gal and Miranker [38] have shown that an optimal choice does exist. Let $M > m > 0$. (In fact it is sufficient to define m and M as the minimum and maximum respectively of the divided differences.) Let the starting interval containing the zero sought be given by

$$x_1 = \frac{M}{(N+1)M+(N-1)m} ,$$

$$x_i = \frac{M+(i-1)(M+m)}{(N+1)M+(N-1)m} , \qquad i=2, 3, \ldots, N .$$

Using N processors the speed-up achieved by the algorithm, assuming that $M+m \gg N(M-m)$, is

$$S_N = 0(\log_c N)$$

with

$$c = (M+m)/(M-m) .$$

Epilogue – Future Trends

Effective algorithms for nonlinear problems are few by comparison with what is available for linear problems. In particular, there are not many procedures for nonlinear optimisation. Some recently developed material in this area is given in Appendix 2.

The methodology of fast transforms (Fourier and number theoretic) is also under current development and has a wide potential for application. Algorithms developed to implement the Fast Fourier Transform were a breakthrough before parallel processing had been realised and since the basis of these transformations is essentially matrix inversion it can be appreciated that parallel versions of the algorithms have great potential. References [39] and [40] provide some background. It may also be mentioned that current developments in software for parallel processing, in particular for the ICL DAP, can be followed by reference to the Documentation List maintained by the DAP Support Unit at Queen Mary College, University of London.

Independently of the technical state of development of computers of MIMD type, which is well behind that of the SIMD family, numerical procedures are needed specifically for the MIMD type. These can, in principle, be implemented on SIMD machines.

Appendix 1

Comparison of performance data of some pipeline, SIMD and MIMD computers

A.0 INTRODUCTION

This material was prepared after the main text had been written and is based on a paper of Schendel and Golm [47]. It enlarges on parts of Chapters 1 and 2 and for this reason alone it seemed worth incorporating in the English translation. The main objective is to provide a comparison, up-to-date at the time of writing, of the performance of some modern high-speed machines and is intended to provide both mathematics and computer-science students with an overview of the current state of the art in computer development, and also to serve as a reference source for specialists. One section has, therefore, been devoted to features of computers that relate to their performance, and another to the most important design characteristics of various machines. A separate section deals with fundamental concepts and, although it overlaps to some extent matters already discussed in Chapter 1, it is felt that this is likely to be quite helpful. The comparisons are based on benchmark tests, the results of which are presented in both tabular and graphic form.

A.1 FUNDAMENTAL IDEAS

A.1.1 SISD, SIMD, MISD, MIMD

The above four abbreviations were introduced by Flynn [4]. They classify machine types according to whether or not they possess two attributes. This may be stated as a classification according to the truth values of the following two Boolean propositions:

(i) At a given moment the machine can process more than one instruction;
(ii) At a given moment the machine can process more than one operand.

The four possible combinations of truth values leads to the following:

00: Single Instruction, Single Data (SISD)
01: Single Instruction, Multiple Data (SIMD)
10: Multiple Instruction, Single Data (MISD)
11: Multiple Instruction, Multiple Data (MIMD)

Flynn's classification helps to indicate the various machine architectures, but it is rather crude.

The SISD category includes the conventional von Neumann machine.

The MISD category is discussed in the literature in various ways. Giloi [41] writes: 'MISD means that a whole stream of instructions is needed to operate on a single data item. The level of organisation would then correspond to that of a von Neumann machine, and since this makes separate classification meaningless, the MISD category is empty'. Other sources interpret MISD as the possibility of applying a number of instructions to a given operand, but this is hardly different from what Giloi has stated and the conclusion that the MISD class is empty must still hold. In passing it is to be noted that the attempt to classify pipeline machines as MISD leads to contradictions.

There remain two categories.

The SIMD category incorporates array processors such as the ICL/DAP, S/9 IAP. In addition, pipeline machines such as the CRAY-1, CYBER-200, CYBER-203, CYBER-205, are usually placed in this category. This does seem correct since, as far as the user is concerned, these machines behave like SIMD machines with instructions for vectors. However, it must be said that, from the physical standpoint, these machines are at any given moment always carrying out the component operations on several operands of an overall operation. Moreover, the forms of algorithm required are different from those needed by a true array processor.

To the MIMD class belong machines with several data processors, and multiple processor systems such as the SM 201, HEP. A range of new problems arises and, in particular, the distribution of tasks to the independent processors, and control and synchronisation, are difficult.

A.1.2 Pipeline processors

To implement the pipeline principle of speeding up arithmetic as done, for example, in the TI ASC machine, the individual operations are, when possible, broken down into a sequence of subsidiary operations, each requiring the same time, which can be carried out separately and independently. Consider as an example floating point addition. The following are the subsidiary operations:

Input
Subtraction of exponents
Decimal point alignment

Mantissa addition
Normalisation
Output

Let t_1 be the time required for a floating point addition and let n_1 be the number of pipeline elements: then a single pipeline element requires time $t_p = t_1/n_1$ to perform its contribution to the whole operation. In processing a single addition no speed increase can be achieved, but if a sequence of additions is to be processed the following occurs. Supposing that after each period of time t_p has elapsed (and so the completion of one subsidiary operation) a new data item becomes available, then the first result emerges from the pipeline sequence after time $n_1 t_p$, and, thereafter, at the end of successive periods of duration t_p, a new result is output. If n_2 data (additions) are to be processed, the n_2th data item enters the pipeline at time $n_1 t_p + n_2 t_p$, and leaves it at time $2n_1 t_p + n_2 t_p = (2n_1 + n_2)t_p$. Thus, for vectors with many components, the speed increase is by a factor of n_1. But the time required to organise the data – called the *start-up time* of the pipeline – is not taken into account here.

There are nevertheless disadvantages to set against the advantage in principle available in the fast processing of long vectors. At the time of writing t_p is of the order of 10 ns. while the access time of an item in working store is of the order of 100 ns. To feed the pipeline with data sufficiently fast it is necessary to use a highly segmented store, in practice many banks of storage elements activated at the right moment. Care must be taken to ensure that the data sequence is correctly stored in the right sequence of storage banks. The start-up time of the pipeline is then very high. Naturally, the pipeline can be fed by special vector registers, as in the CRAY-1, but then there arise problems of transferring data between the registers and the store. Since start-up times cannot be avoided when scalars are processed on a pipeline machine experience has for a long time been discouraging in such applications (STAR-100, CYBER-200).

As well as pipelining arithmetic operations, the pipelining of the instruction stream can be considered. That is to say, the execution of a new instruction begins before the execution of its predecessor has been completed. This is by no means a typical feature of modern computers although it is a facility encountered on certain microprocessors – the Z-80 is an example – but the processing of interrupts then presents some difficulties.

A.1.3 Array processors

The reduction in price of digital circuitry made it feasible to contemplate using arrays, or fields, of processors instead of just one, as had been the case. In an array processor each component processor carries out the same instructions and thus we are dealing with an SIMD machine. The processors are usually arranged in matrix form, as in the ICL/DAP, and each processor has its own store and is connected with the others through a suitable network. Variants exist in which each processor is connected to different storage blocks. The diagrams illustrate two possibilities.

Connection of n processor–storage pairs

```
M0       M1       M2       ...      Mn
P0       P1       P2       ...      Pn
|        |        |        |        |
-------------------------------------------
            Connecting network
-------------------------------------------
```

Connection of n processors with n storage modules

```
P0       P1       P2       ...      Pn
|        |        |        |        |
-------------------------------------------
            Connecting network
-------------------------------------------
|        |        |        |        |
M0       M1       M2       ...      Mn
```

The connecting network must fulfil the role of a bijective connecting function f: $0, 1, \ldots, n-1 \rightarrow 0, 1, \ldots, n-1$, under program control. In practice, it is more usual to limit the scope of the possible connecting function, in particular to use a neighbouring lattice point procedure where each processor can communicate only with its immediate neighbours.

Array processors are often added to normal scalar processors, the case with the ICL/DAP. Since array processors use exactly the same time to carry out a scalar operation as an operation on a matrix or vector they are to be considered slow for scalar operation. On the ICL/DAP combination all scalar operations are performed on the conventional scalar machine, matrix and vector operations being naturally reserved for the DAP.

A.2 PERFORMANCE CHARACTERISTICS OF COMPUTERS

A.2.1 Mflops

Mflops (pronounce 'megaflops') is the accepted abbreviation for 'million floating point operations per second'. Its use, however, as an index of performance may be misleading. For example, manufacturers sometimes proceed as follows: take the time of the fastest floating point operation on a machine and describe its reciprocal as the megaflop rate in the prospectus. The technique is also favoured

of selecting a program which, by means of clever tricks, performs many floating point operations in a short time, and using this as an indicator of the Mflop rate.

Measures calculated in this way are neither helpful nor relevant for a potential purchaser. A more objective procedure is required to arrive at verifiable statements of the Mflop rate. The usual method is to run *benchmark tests,* that is, programs typical of the application envisaged, and to calculate the Mflop rate from carefully controlled and timed repetitions on the machine under evaluation. In this connection the *throughput,* that is, the amount of work processed by the machine when several tasks are in concurrent operation, is of importance and not to be neglected. Since many practical problems do not involve carrying out the largest possible number of floating point operations in the least time (in numerically intensive programs this may not be the case) the Mflop rate must be considered in conjunction with other relevant measures, of which throughput is an example.

A.2.2 Main store

The *main store* of a computer is the store to which the *central processing unit* (CPU) has direct access. It is not to be confused with *virtual store* to which logical access must also be available, but which, being copied to and from main store, entails slower access.

Next, the *word length* of the store is of interest. A *word* is the shortest addressable unit in the store. In many machines floating point numbers are represented in one word and, in this case, the longer the word, the greater the accuracy of the machine. It is thus in problems requiring high accuracy that a long word is of greatest advantage. In commercial problems this is less the case. The size of the store is an important feature of a machine. Of what avail is a fast computer if a given problem cannot be solved because of limited storage capacity? In machines which use the virtual store concept it must be borne in mind that this entails slower access. The *transfer rate* of a store is the number of bits tansferred in unit time. It indicates the speed of store and is also used to characterise the performance of external storage media such as disk systems, though in this connection head movement times and other imponderables are factors requiring careful consideration.

A.2.3 Mass storage

By *mass storage* is understood peripheral store to which the processor does not have direct access. The connection is made through special control devices. The usual types of mass store are disk drives and magnetic tape stations. Since the mass store is not connected directly to the processor its capacity is to all intents unlimited (except by cost!). The transfer rate of data is again of great importance. If too low, a bottleneck may be created, and thus a weak point in the system. For this reason an extra mass store was developed for the CRAY series, which consists entirely of special integrated storage switching circuits. This endows the CRAY-1 with a significantly improved throughput. Fast mass store is of particular importance in machines employing the virtual storage concept.

A.2.4 Benchmark tests

Benchmark tests are an important aid for determining the performance of a machine. Such tests consist in running a package of programs representative of the proposed use of the machine. By running the same benchmark programs on different machines it is possible to make a reliable comparison between them based on the results which should, as emphasised above, include *throughput* as well as running time. Frequently, however, this is neglected, a great mistake, for a machine even three times faster than another but yet on average utilising only 20 per cent of its capacity is, as a rule, inferior to the slower machine if this is, say, 90 per cent occupied on average.

A.2.5 Throughput

High performance machines today possess *multiprogramming* operating systems which means the capability of processing several programs simultaneously. The available time is divided among the different programs and this is called *time-sharing.* The processor leaves a given program after the elapse of a preordained time or when a time-intensive input/output operation is called for, and then passes to another program. It could happen that, at a given moment, the processor is idle, for example because all the programs being processed call for input/output operations. The summation of such times is called *idle time* and is a factor which degrades the performance of the machine.

Throughput is a measure of idle time. The ideal is 100 per cent throughput (no idle time) meaning that the processor is in continual operation. A lower throughput tends to indicate that the system has been slowed down through mass storage transfer. In benchmark tests it often happens that only the CPU time involved in running the program suite is given, and this is not by any means enough for a thorough appraisal of a machine.

A.2.6 Software

The importance of software as a factor influencing performance is often overlooked. Many modern, fast machines offer excellent hardware, but relatively little in the way of accompanying software. The offer made by a manufacturer usually includes FORTRAN IV or V compilers with some extensions that make use of the particular hardware that is a feature of the machine, or perhaps indications how to organise programs so as to optimise the compiler (CRAY-1). Sometimes, at additional cost, it is possible to acquire a FORTRAN or other program library. PASCAL is a feature of some machines. ALGOL seems to be less popular nowadays.

It is surprising how few machines possess compilers for languages such as APL and ALGOL-68 which support at least the concept of parallel algorithms and matrix and vector operations. Clearly demand is not great, or has not been recognised by the manufacturers. Such facilities could, however, greatly enhance a machine and this must at least be recognised by Control Data Corporation, the only manufacturer (other than IBM) known to us at the time of writing to offer APL – for the CYBER-205. It must be concluded that at present software provision leaves something to be desired.

A.2.7 Prices

Many manufacturers appear to hold to the opinion that price is unimportant. The author, who does not share this opinion, attempted to obtain some price data for inclusion in this review but it turned out to be very difficult. One reason is that the market is highly competitive, and secretive as a consequence. Manufacturers and clients enter into complex n-person games during the procurement process for a new computer. The payoffs are variable and, at all costs to be withheld from competitors.

A.3 BRIEF DESCRIPTION OF MACHINES

A.3.1 CRAY-1, CRAY-1S, CRAY X-MP

In 1972 Seymour Cray founded the firm Cray Research. Previously he had worked for the Control Data Corporation and had participated in the development of the CDC 6600 and 7600. In April 1976, work on the initial model was completed and the first CRAY-1 went into service at the Los Alamos Scientific Laboratory. The S-model has been on the market since 1980 and differs from its predecessor in possessing enhanced main store and improved input–output facilities. In 1982 the CRAY X-MP was proposed. The idea was to use more highly integrated components in order to install two improved 1/S processors in one housing. It is thus a twin-processor machine. Nevertheless it is not an MIMD machine but rather two high-performance machines which share a common database and peripherals. The unusual construction of CRAY machines should be mentioned. In order to achieve the highest possible speed of transmission of signals within the machine all cables are restricted to a length of less than one metre and a special housing design is employed. The CRAY is a cylinder of height 1.905 metres and diameter 2.87 metres up to a height of 0.48 metres (a convenient sitting height), and of 1.372 metres from there up. Computation always takes place in the vector register. The programming language is ANSI FORTRAN V and the compiler carries out to a limited extent the process of vectorisation. These vectorisations are, however, essentially limited to DO-loops which arise in connection with array operations. A more suitable programming language would render vectorisation unnecessary but the portability of the very many FORTRAN programs already in existence would suffer. The exceptionally short compile time is a feature to be noted on the CRAY machines, that used for FORTRAN being about ten times faster than that of the CYBER-205.

Some CRAY-1/CRAY-1S data follow:

Word length: 64 bits
Store size: 2–8 Mbytes/2–32 Mbytes
Clock time: 12.5 ns
Store cycle time: 50 ns
Register: 8 to 64 words
Input/output transfer rate: 20,000 Mbytes/s/10,000 Mbytes/s
1–4 I/O processors

The performance, subject to previously mentioned reservations, is about 80 Mflops.

In the CRAY X-MP the store cycle time has been decreased to 9.5 ns, the number of vector registers increased to 32, and the I/O transfer rate increased to 200,000 Mbytes/sec. The storage access bandwidth has been increased from one word per cycle to two, or possibly more, and thus an inherent bottleneck has been removed. Another problem was that throughput in the CRAY-1 and -1-S was limited by the speed of disk storage, and this has been overcome by the introduction of fast integrated circuit store with capacities of 64, 128, and 256 Mbytes. The X-MP has several times the performance in Mflops of the 1-S, namely between 500 and 600.

A.3.2 STAR-100, CYBER-200, CYBER-203, CYBER-205

The STAR-100 was the first large machine to be conceived for vector operation on the pipeline principle. It is a product of the early 1970s and was realised with the technology then available. It consists of highly interleaved core store of 4 Mbyte capacity, two pipeline processors and a series of peripheral processors for input/output, operating system functions and data management. The primary design objective was the highest possible processing speed for vector operations. The basic clock time is 40 ns, but the start-up time of the pipeline is very high – of the order of 3000 ns. The machine was not produced in series: instead CDC concentrated on modernising the hardware, and the result was the appearance of the CYBER-200 series at the beginning of 1979.

The programming language of the CYBER-200 series is a dialect of FORTRAN IV containing extensions for vector and matrix instructions. The compiler possesses to a limited extent the capability of vectorising and optimising standard programs. In addition to FORTRAN, APL is available.

A powerful scalar unit was incorporated in the CYBER-203 which has enhanced the speed of scalar operations. In the CYBER-205 the pipeline has been significantly improved.

In the CYBER-200 series computation is always carried out in the memory and not, as in the CRAY, in the register. For this reason it is not unusual to experience high start-up times for vector operations. With the CYBER-203 this could be 159 cycles at 40 ns per cycle, that is between 6 and 6.5 μs. The improvement achieved by the CYBER-205 is 51 cycles at 20 ns, giving about 1 μs.

Some data for the CYBER-203/205 follow:

Pipeline elements: 2/2 or 4
Word length: 64 bits
Store: 4–16 Mbytes/up to 32 Mbytes
Clock time: 40 ns/20 ns
Store cycle time: 80 ns/40 ns(?)
Input/output transfer rate: 60 Mbyte/s/?

In the processing of 32-bit half-words the performance is of the order of 100 Mflops/400 Mflops. This is halved when whole words are processed, but it must be noted that such performance is attainable only when long vectors are being processed.

A.3.3 DENELCOR HEP (heterogeneous element processor)

In 1973 the firm DENELCOR Inc. began the development of a digital computer which was intended to combine the advantages of parallelism and speed obtainable with analogue machines, with those of programmability, flexibility and reliability associated with digital machines. This development led to the construction of a prototype which was supported by the US Army Ballistic Research Laboratory. In spite of initial technical problems a further contract with BRL enabled an improved model to be planned.

The HEP, introduced in 1980, is an MIMD machine using shared resources. A maximum of 16 processors can be connected with a maximum of 128 data storage modules each of which has the capacity of up to 1 Mword of 64 bits. In each processor up to 50 user processes can be active. Eight completely independent processes can be carried out simultaneously and this can be thought of as the operation of a pipeline with eight components. A new instruction can be brought into the pipeline every 100 ns. If a process requires access to the store it is detached from the instruction stream and placed in a queue which controls access through the connecting network. Only when access to store has been gained is the process again placed in communication with the instruction stream. Processes communicate by means of common protected storage regions. Synchronisation is controlled by a producer–consumer protocol. Each instruction tests whether its source is occupied and its destination empty. If one of these tests is false the process must wait. If both tests give a true result the source is set to empty and the destination set to reserved. Then the allocation of data takes place and the destination sat to occupied.

Each processor can execute ten million instructions per second. Since with array operations index calculations are likely to be met a performance figure of about 3 Mflops per processor is a reasonable estimate. The connecting network has a transfer capacity of ten million 64-bit words per second per connecting circuit, to a maximum of 16.

According to the DENELCOR engineers, higher programming languages present difficulties because of the unusual architecture of the HEP. ALGOL-68 might have been worth consideration, but seems little known or used in the USA and, as a result, an extended version of FORTRAN was settled on. It is strange that no further proposals seem to have been made.

A *process* is generated by means of the CREATE command, analogous to the familiar CALL command. In addition to the normal variables there exist asynchronous variables which possess condition indicators as well as values. The result is that it is difficult to organise a program optimally for the HEP taking proper account of the overhead associated with the CREATE function.

Because of its special architecture it is difficult to make a fair comparison between HEP data and those of more orthodox machines. Nevertheless we present the following:

Word length: 64 bits
Clock period: 100 ns
Store size: Up to 128 units of 1 Mword
Input/output transfer rate: Up to 80 Mbytes per second
Number of processors (Process Execution Modules or PEMs) 1–16.
Program memory per processor: 32k or 1 Mword
General registers per processor: 2048
Special registers per processor: 4096
Maximum user process status words per processor: 64
Maximum task status words per processor: 8

Since the machines are extremely flexible and modular it is relatively easy to make additions, but this increases the cost in proportion.

A.3.4 ICL/DAP

The DAP (Distributed Array Processor) consists of an array of 64 × 64 one-bit processors instructed to perform the same operation by a single master processor, the Master Control Unit. Each processor can communicate with four neighbours. The DAP is an appendage to an ICL-2900 computer which deals with the scalar parts of a program and the programmer has the task of deciding how to divide his program between the 2900 and the DAP. The programming language is an extended FORTRAN with special facilities for array processing.

The 2900 manages the DAP as a 2 Mbyte additional store with 4 Kbits per processor element. Significant improvements in performance are obtainable when a program can be broken down into a large number of subprograms, decompositions which are possible for logical, character and picture processing.

A.3.5 Hitachi S9/IAP

The Hitach S9/IAP is a combination of a S9 scalar machine and an Integrated Array Processor (the IAP). The scalar computer has an IBM-compatible instruction set. The equivalent machine for the Japanese market, where IBM compatibility is not necessarily imperative, is the M-280H which has another instruction set. The scalar computer possesses a high-speed arithmetic unit (HSA) as standard.

The IAP represents another example of processing on the pipeline principle. The IAP shares HSA and store with the S9. As in the CYBER-205, the HSA vector flow is driven from store to store. An advantage of this is that HSA and normal arithmetic can operate concurrently.

The pipeline startup time is lower than that of the CYBER-205, but a new result is obtainable only after 60 ns in contrast with 10 ns for the CYBER. The programming language is standard FORTRAN 77 and, as with the CRAY, the compiler vectorises the innermost DO-loops according to certain rules. Thus it is possible to enjoy the advantages of IAP without having to rewrite programs. Significant speed gains are nevertheless to be achieved by changing double loops to single ones and reorganising matrix operations into vector ones.

Since the S9 is available with both 8-fold and 16-fold interleaved memory the results of benchmark tests must naturally be interpreted in this light.

Some data for the S9/IAP follow:

Word length: 64 bit (double precision)
Memory size: 8–32 Mbyte
Clock time: 30 ns
Input/output transfer rate: 3 Mbyte per second per channel (with a maximum of 90 Mbyte per second with 32 channels)

Hitachi has announced the supercomputer S-810/20 for the third quarter of 1983, and a reduced version (S-810/10) is also to become available. A performance of up to 630 Mflops is expected. The machine is to consist of an IBM-370-compatible scalar machine and a pipeline processor having 32 registers with 256 8-byte words. The main store can be augmented up to 256 Mbyte. Up to 32 I/O channels each with 96 Mbyte per second are furnished. The machine can be coupled with an S9 and share its peripherals.

To solve the problem of memory speed, particularly troublesome for virtual memory management, mass storage consisting of fixed integrated circuits is used, as with the CRAY-X-MP. It has a maximum capacity of 1 Gbyte (that is 1024 Mbyte) and a transfer rate of 1 Gbyte per second; it is thus 300 times faster than the fastest Hitachi disk drive.

Programming of the S-810/10 will be in standard FORTRAN 77 and a special mathematical library is said to be ready. Unfortunately, this is the total of the information available at the time of writing.

A.4 RESULTS OF BENCHMARK TESTS

A.4.1 Introductory remarks

In what follows the results of certain benchmark tests will be described. The principal source is the report of Müller-Wichards and Gentzsch [42] published by the DVFLR (Deutsche Versuchs- und Forschungsanstalt für Luft- und Raumfahrt). Two kinds of test were run:

(i) timings of a set of basic computational operations on scalars and vectors;
(ii) timings of a set of programs typical of day-to-day work at the DVFLR.

As far as possible the tables give Mflops, but, since there was no indication in the DVFLR programs of the number of operations involved, a different representation of performance has been chosen. Let the fastest time to run a given program be denoted by W_f. The corresponding machine is allocated a performance indicator of 100%. Let the time to run the same program on another machine be W_a. That machine's performance is then given as $100 W_a/W_f$ %. In one case (Table A.6) the actual timings are given since they allow interesting conclusions to be drawn.

The following machines were tested as indicated.

CYBER-205	CDC-Data center, Arden Hills
CRAY-1S	CRAY-Research, Chippewa Falls
CRAY-1	IPP, Garching
DENELCOR-HEP	DENELCOR-Works, Aurora
Hitachi S9/IAP	Hitachi-Works, Kanagawa
ICL-2960/DAP	Queen Mary College, University of London

For comparison the following scalar machines were used:

AMDAHL-470/V6	DVFLR, Oberpfaffenhofen
IBM-3081 D	DVFLR, Oberpfaffenhofen

The following special remarks should be noted:

(i) A two-pipe version of the CYBER-205 was used.
(ii) The timings of the IBM-3081D refer to a single processor (only one processor can work on one task.
(iii) Since the timings of the CRAY-1 with 16 storage banks were hardly different from those of the CRAY-1S with 8 storage banks only the timings of the CRAY-1 are given.
(iv) Large arrays were stored in the Large Core Memory (LCM) of the CYBER. This caused a reduction in performance.
(v) The 8- and 16 storage bank versions of the Hitachi-S9/IAP showed significant differences. Unless otherwise stated the better times obtained with the latter version are given.
(vi) The HEP was a single processor version.
(vii) The benchmarks ran on the IBM-3081D and AMDAHL-470/V6 using the HE-FORTRAN IV compiler of the MVS-SE operating system. For the

CYBER the FORTRAN IV compiler of the operating system SCOPE 2.1 was employed.

A.4.2 Results of tests of fundamental kernels

The fundamental kernels referred to constitute a set of 10 important computation types. These are now explained.

Kernel no.	Definition	No. of operations
1	$x+y$	1
2	$x*y$	1
3	x/y	1
4	$(x-y)*(x+y)$	3
5	$\langle x, y\rangle$	2
6	$5*(x/z+y*z)$	4
7	$P_9(x)$	20
8	x_i-x_{i-1}	1
9	$x_{i+1}-2x_i+x_{i-1}$	3
10	$y-B*x$	2

Let x and y be N-dimensionsal vectors $x=(x_1, x_2, \ldots, x_N)^T$, $y=(y_1, y_2, \ldots, y_N)^T$. Then Kernel 1 is the formation of the vector $z=(z_1, z_2, \ldots, z_N)^T$ where $z_i=x_i+y_i$, $i=1, 2, \ldots, N$. This definition of a new vector z can be extended similarly to explain Kernels 2, 3, 4, 5, 6 and 10. Note that $\langle x, y\rangle$ in Kernel 5 denotes the scalar product $\Sigma x_i y_i$. In Kernel 10, B is a scalar constant. In Kernel 7, $P_9(x)$ denotes the evaluation of a polynomial of degree 9 at the points $x_i (i=1, 2, \ldots, N)$ by Horner's Scheme. For Kernel 8 the first difference vector $z=(x_1, x_2-x_1, \ldots, x_N-x_{N-1})$ is formed. Kernel 9 entails second differences.

Table A.1 – Scalar kernels; vectorisation facility switched off. (Vector length N=100; performance in Mflops)

Kernel	AMDAHL	IBM	Hitachi	CRAY	CYBER-205	CYBER-76	HEP
1	1.04	1.51	3.65	2.62	2.73	2.78	0.060
2	0.75	1.13	3.00	2.54	2.91	3.23	0.060
3	0.33	0.60	0.98	1.39	0.84	1.39	0.057
4	1.49	2.69	5.84	6.22	7.77	5.66	0.133
5	1.61	2.16	5.49	4.63	5.80	4.76	0.116
6	0.82	1.55	3.09	4.47	3.06	3.15	0.171
7	1.19	1.94	5.08	4.03	5.77	2.67	0.103
8	0.81	1.49	3.70	2.65	2.89	2.86	0.061
9	1.36	2.39	6.82	5.89	7.19	5.26	0.159
10	–	1.90	5.46	4.50	6.13	3.45	0.110
Average	1.04	1.74	4.31	3.89	4.51	3.52	0.103

Table A.2 – Vector kernels; autovectorisation facility switched on. (Vector length N=100; performance in Mflops)

Kernel	AMDAHL	IBM	Hitachi	CRAY	CYBER-205	HEP (P=20)	HEP (P=10)
1	1.04	1.51	11.49	18.87	43.48	0.141	0.156
2	0.75	1.13	11.36	18.87	50.00	0.142	0.156
3	0.33	0.60	1.21	10.10	12.99	0.141	0.161
4	0.149	2.69	10.60	43.48	50.00	0.419	0.452
5	1.61	2.16	25.00	32.26	41.67	0.156	0.163
6	0.82	1.55	3.71	30.30	29.20	0.527	0.610
7	1.19	1.94	13.05	35.59	59.52	0.182	0.283
8	0.81	1.49	9.80	18.87	45.45	0.144	0.157
9	1.36	2.39	11.32	37.50	47.62	0.420	0.469
10	–	1.90	19.61	27.03	76.92	0.283	0.317
Average	1.04	1.74	11.70	27.30	45.70	0.256	0.292
Speed-up	1.0	1.0	2.7	7.0	10.1	2.5	2.8

The final row in Table A.2 is the speed-up ratio obtained by dividing the average of Table A.2 by that of Table A.1.

The next Table A.3 is the same as Table A.2 with N=1000.

Table A.3 – Vector kernels with autovectorisation. (Vector length N=1000; performance in Mflops)

Kernel	AMDAHL	IBM	Hitachi	CRAY	CYBER-205	HEP (P=20)	HEP (P=10)
1	1.07	1.56	13.91	22.94	87.72	0.198	0.237
2	0.74	1.12	15.36	22.73	90.09	0.198	0.237
3	0.31	0.60	1.23	11.34	15.65	0.218	0.236
4	1.66	2.68	10.83	51.11	89.82	0.620	0.683
5	1.35	2.16	32.00	66.45	88.11	0.173	0.185
6	0.80	1.53	3.95	33.53	41.19	0.825	0.898
7	1.27	1.92	17.00	43.93	160.90	0.415	0.453
8	1.09	1.46	12.97	22.60	88.50	0.200	0.233
9	1.78	2.32	12.05	42.08	90.36	0.614	0.698
10	–	1.89	26.11	42.28	178.57	0.366	0.463
Average	1.12	1.72	14.50	35.90	93.10	0.383	0.432
Speed-up	1.0	1.0	3.4	9.2	20.6	3.5	4.2

Note that the table corresponding to Table A.1 is not shown in this case.

A.4.3 Results of the benchmark programs

Timings were carried out using typical DVFLR programs mainly concerned with the solution of boundary value problems in fluid dynamics. The programs were run both in unmodified form and also in a form adapted specially to the individual machine under test. When the unmodified form was used the autovectorisation facility where appropriate was switched on. In each case the HEP used specially adapted programs. Standard assembler routines were embodied into some of the adapted programs and this led to significant improvements in performance.

Table A.4 gives results obtained with the set of unmodified programs. These, briefly, are as follows:

WALL: Arises from investigations into shock-boundary layer interaction in low temperature nitrogen
MGO1: Solution of Helmholtz equation in a general domain.
TR3: Second order discretisation for solution of Dirichlet problem in rectangle.
MAD: Solution of Poisson equation in rectangle
E3D: Three-dimensional non-viscous supersonic flow about blunt body.

Table A.4 – Unmodified programs with autovectorisation; performance as percentage of best performer

Program	AMDAHL	IBM	Hitachi	CRAY	CYBER-205
WALL	15.7	23.2	63.1	100.0	60.8
MGO1	25.0	29.1	67.3	99.4	100.0
TR3	15.0	22.6	67.4	100.0	94.3
MAD	11.4	19.3	66.7	100.0	41.1
E3D	13.2	20.1	55.5	100.0	80.6
Average	16.2	22.9	64.0	99.9	75.3

The adapted programs whose performance is listed in Table A.5 were briefly as follows:

MHD 12 and 22: Magnetohydrodynamic investigations
MGOO: Again Helmholtz equation in a rectangle
EPS 4 and 7: Integral equations
GEL 100 and 200: Gauss elimination with non-sparse matrices
E2D: As E3D in two dimensions
RELAX 15 and 63: Gauss–Seidel relaxation procedures with sparse matrices.

Table A.5 – Adapted programs; perfc rmance as percentage of best

Program	AMDAHL	IBM	Hitachi	CRAY	CYBER-205	HEP (Ser)	HEP (Par)
MHD 12	4.1	6.9	33.3	75.0	100.0	0.4	2.1
MHD 22	2.5	3.7	22.6	62.3	100.0	–	–
MG 00	4.4	7.3	25.3	94.2	100.0	–	–
EPS 4	9.9	15.2	61.1	89.7	100.0	0.7	5.6
EPS 7	8.8	12.6	53.8	88.0	100.0	0.6	5.3
GEL 100	1.9	3.1	26.2	100.0†	64.7	0.1	1.0
GEL 200	0.9	1.5	15.6	100.0†	52.5	0.1	0.5
E2D	13.1	22.5	–	100.0	70.7	–	–
RELAX 15	2.3	3.8	7.4	31.3	100.0	–	–
RELAX 63	1.1	2.0	6.3	33.3	100.0	–	–
Average	4.9	7.9	27.9	77.4	88.9	0.4	2.9

†Uses standard assembler routines.

The influence of vector length was investigated for the MHD programs. The following results were obtained.

Table A.6 – Influence of vector length N in MHD program; time for 1000 iterations (seconds)

N	AMDAHL	IBM	Hitachi	CRAY	CYBER-205	DAP†
10	4.4	2.6	0.54	0.24	0.18	11.0
20	15.4	10.2	1.68	0.61	0.38	11.0
30	36.3	22.4	3.47	1.17	0.75	11.0
40	63.8	39.0	5.94	1.83	1.22	11.0
50	100.1	58.4	9.09	2.75	1.83	11.0
60	131.9	83.0	13.06	3.66	2.56	11.0

†With 32-bit arithmetic.

Table A.7 – Influence of vector length N in MHD program: Part 2; performance expressed as percentage of best

N	AMDAHL	IBM	Hitachi	CRAY	CYBER-205	DAP†
10	4.1	6.9	33.3	75.0	100.0	1.6
20	2.5	3.7	22.6	62.3	100.0	3.5
30	2.1	3.3	21.6	64.1	100.0	6.8
40	1.9	3.1	20.5	66.7	100.0	11.1
50	1.8	3.1	20.1	66.5	100.0	16.6
60	1.9	3.1	19.6	69.6	100.0	23.3

†With 32-bit arithmetic.

A.4.5 Discussion of results

Since all the important data have now been given the reader can very largely draw his own conclusions. It is nevertheless worth while drawing attention to some points.

The improvement in performance attainable with modern machine architecture is considerable. This is particularly noticeable when comparing IBM's most modern machine (at the time of writing the 3081D) with the CYBER-205 and CRAY-1. Even if the throughput of the IBM can be doubled by the use of two processors it remains inferior to the other machines by an order of magnitude. The CYBER-205 performs significantly better when specially adapted programs are used than the CRAY-1, and somewhat worse with unmodified programs, but it can be regarded as 'winning the competition'. It should nevertheless be remembered that the CYBER was brand new while the CRAY was almost out of date. The CRAY-X-MP would have provided a more correct comparison since each processor of this twin processor machine has roughly two and a half times the power of the CRAY-1, and in this case the CRAY would have 'won'. It is unfortunate that the new Hitachi could not be considered for Hitachi had announced that it would have ten times the performance of the S9/IAP and it would therefore most probably have ranked among the leaders. Incidentally it bears an astonishing resemblance to the CRAY except that everything is bigger and faster. The S9/IAP also performed exceptionally well and taking into account its favourable price must be considered a most attractive prospect. For many users its IBM compatibility will certainly constitute an advantage. Too little information was available for the DAP and a clear picture cannot be formed. Unfortunately the new Fujitsu could not be considered because of lack of documentation and test reports.

Appendix 2

Some procedures for nonlinear optimisation

During the past ten years increased efforts have been made to construct parallel algorithms for nonlinear optimisation. The procedures described here are due to Sloboda [43] and Chazan and Miranker [44], and do not require the derivatives of the function to be minimised. They are based on a serial conjugate Gram–Schmidt procedure. The underlying notion is to calculate the function values in parallel in order to reduce the time needed by the minimisation procedure. The following summary is based on studies by Schendel and Schyska [48], [49].

Most serial procedures for the minimisation of a function f of n real variables generate iteratively a sequence of n-vectors x_k (k=2, 3, . . .) of real components starting from a suitable initial vector x_1. Each step of the iteration consists of two parts:

Step 1: Calculate according to prescribed rules a search direction p_k, a real n-vector, where the subscript denotes the iteration currently in progress.

Step 2: Determine by means of a search along a line (linear search) a step length $a_k \in \mathbb{R}$ such that, at least approximately,

$$f(x_k + a_k p_k) = \min_a f(x_k + a p_k) \quad ,$$

and then set

$$x_{k+1} = x_k + a_k p_k \quad .$$

Individual procedures differ essentially only in the way the search direction is chosen. Conjugate direction procedures are such that for a quadratic function the global minimum is reached using exact linear search in not more than n steps, where n is the number of variables in the problem.

For the purposes of this introductory description of the Gram–Schmidt procedure it will be assumed that the function to be minimised $f\colon \mathbb{R}^n \to \mathbb{R}$ is the quadratic form

$$f(x) = \tfrac{1}{2} x^{\mathrm{T}} A x - b^{\mathrm{T}} x + c \quad , \qquad \text{(A.2.1)}$$

where A is a symmetric $n \times n$ matrix of real elements, b is a real n-vector, c is a real number. The unique global minimum point (real n-vector) x^* satisfies

$$x^* = A^{-1} b \quad . \tag{A.2.2}$$

To save repetition it may be assumed that all matrices and vectors and scalars have real components, unless otherwise stated. Moreover it will be assumed that an n-dimensional problem is being handled.

An important property of quadratic functions is the uniqueness of minimum points on lines. In this connection we enunciate the following

Lemma

The minimum point x_2 of (A.2.1) on the line

$$L := \{x \mid x = x_1 + \alpha p_1, \alpha \in \mathbb{R}\}$$

passing through x_1 in the direction p_1 is given uniquely by

$$x_2 = x_1 + a_1 p_1$$

where the step length a_1 is given by

$$a_1 = -\frac{p_1^{\mathrm{T}} (A x_1 - b)}{p_1^{\mathrm{T}} A p_1} \quad .$$

Next we recall the concept of pairwise conjugacy of vectors with respect to a matrix. This is a generalisation of orthogonality. Thus, the orthogonalisation of the Gram–Schmidt procedure can be adapted to the construction of n pairwise conjugate vectors $\hat{p}_1, \ldots, \hat{p}_n$ with respect to the matrix A from n linearly independent vectors $u_1, \ldots, u_n$. Using the rule

$$\begin{aligned}
\hat{p}_1 &= u_1, \\
\hat{p}_2 &= u_2 - b_{21}\hat{p}_1, \\
&\vdots \\
\hat{p}_n &= u_n - b_{n1}\hat{p}_1 - b_{n2}\hat{p}_2 - \ldots - b_{n,n-1}\hat{p}_{n-1} \quad ,
\end{aligned}$$

we have, from the conjugacy requirement

$$\hat{p}_i^{\mathrm{T}} A \hat{p}_k = 0 \qquad (k = 2, \ldots, n; i = 1, 2, \ldots, k-1) \quad ,$$

the result

$$b_{ki} = \frac{u_k^{\mathrm{T}} A \hat{p}_i}{\hat{p}_i^{\mathrm{T}} A \hat{p}_i} \qquad (k = 2, \ldots, n; i = 1, \ldots, k-1)$$

for the conjugacy coefficients.

This leads to

Algorithm 1
An initial value (point)

$$x_1 \in \mathbb{R}^n \tag{A.2.3}$$

is given, together with n linearly independent vectors

$$u_1, u_2, \ldots, u_n \in \mathbb{R}^n. \tag{A.2.4}$$

The kth step of the iteration is then broken down into:

Step 1: Calculate the kth search direction $\hat{p}_k$ from

$$\hat{p}_k = u_k - \sum_{i=1}^{k-1} b_{ki} \hat{p}_i \tag{A.2.5}$$

with

$$b_{ki} = \frac{u_k^T A \hat{p}_i}{\hat{p}_i^T A \hat{p}_i} \quad , \quad i = 1, \ldots, k-1 . \tag{A.2.6}$$

Step 2: Determine by one line search starting from the point x_k the step length

$$a_k \in \mathbb{R} \tag{A.2.7}$$

and put

$$x_{k+1} = x_k + a_k \hat{p}_k \quad . \tag{A.2.8}$$

With exact line search the algorithm terminates at the global minimum point of (A.2.1), that is

$$x_{n+1} = x^* \quad , \tag{A.2.9}$$

from the pairwise conjugacy of the search directions $\hat{p}_1, \ldots, \hat{p}_n$ with respect to the matrix A (see, for example, [45]).

Formulae (A.2.5) for the computation of the search directions form a linear recurrence $R\langle n, n-1\rangle$ of order $n-1$ with n variables. Several parallel algorithms have been developed for the solution of such recurrences. See Chapter 3 of this book, for more information. Particularly suitable for $R\langle n, n-1\rangle$ is the Column-Sweep Algorithm and it can thus be used to compute the search directions $\hat{p}_1, \ldots, \hat{p}_n$.

Figure 1 shows the procedure diagrammatically.

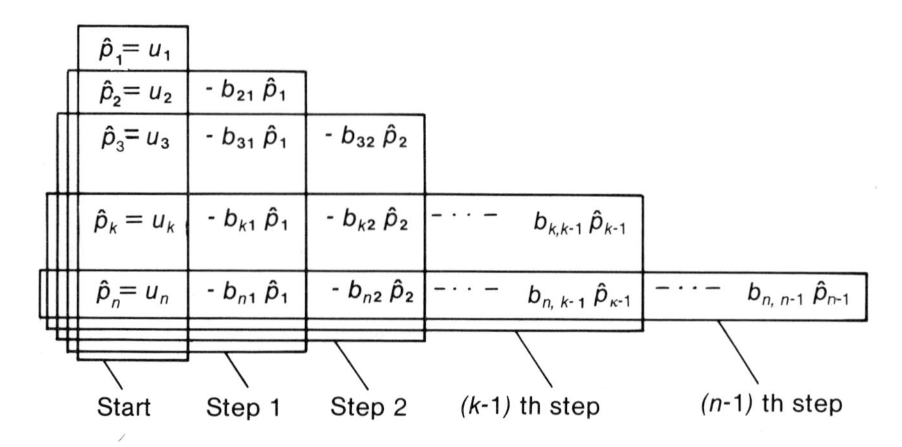

Fig. 1 – Illustrating the procedure for the parallel computation of (A.2.5).

To begin, the following assignment is made in parallel for i=1, . . . , n:

$$\hat{p}_i^{(0)} = u_i \quad . \tag{A.2.10}$$

Then the coefficients

$$b_{ik} = \frac{u_i^{\mathrm{T}} A \hat{p}_k}{\hat{p}_k^{\mathrm{T}} A \hat{p}_k} = \frac{\hat{p}_i^{(k-1)^{\mathrm{T}}} A \hat{p}_k^{(k-1)}}{\hat{p}_k^{(k-1)^{\mathrm{T}}} A \hat{p}_k^{(k-1)}} \tag{A.2.11}$$

are computed one after the other for $k = 1, \ldots, (n-1)$, and at each step this is done in parallel for $i = k+1, \ldots, n$. Following this, the vectors

$$\hat{p}_i^{(k)} = \hat{p}_i^{(k-1)} - b_{ik}\hat{p}_k^{(k-1)} \tag{A.2.12}$$

are calculated, again in parallel, for $i = k+1, \ldots, n$. From (A.2.5) and (A.2.6) it follows that for all $k = 1, \ldots, n$

$$\hat{p}_k^{(k-1)} = \hat{p}_k \quad . \tag{A.2.13}$$

The following Lemma from [45] explains how conjugate directions can be calculated without explicit knowledge of the derivatives of the function to be minimised.

Lemma

Let f be a quadratic function of form (A.2.1) and let $x_2, \bar{x}_2$ be minimum points of f on the lines

$$L = \{x \,|\, x = x_1 + \alpha p_1\} \tag{A.2.14}$$

$$\bar{L} = \{x \,|\, x = \bar{x}_1 + \alpha p_1\} \tag{A.2.15}$$

through x_1 and $\bar{x}_1$ respectively with the same direction vector p_1. Then

$$(x_2 - \bar{x}_2)^T A p_1 = 0 \quad . \tag{A.2.16}$$

By suitable choice of starting points for line searches the search directions of the conjugate Gram–Schmidt procedure can also be calculated in terms of differences of minimum points of f on lines instead of using the formulae (A.2.5), (A.2.6), (A.2.11), (A.2.12). If, moreover, the line searches for the calculation of these minimum points are carried out in parallel we obtain the parallel procedure of Sloboda which will be described next. In order to compare the two processes the function f will continue to be of form (A.2.1).

Algorithm 2
Given an arbitrary starting point

$$x_1^{(0)} \in \mathbb{R}^n \tag{A.2.17}$$

and n linearly independent vectors

$$u_1, \ldots, u_n \in \mathbb{R}^n. \tag{A.2.18}$$

Start: In parallel set

$$x_1^{(i)} = x_1^{(0)} + u_i \qquad (i = 1, \ldots, n) \quad . \tag{A.2.19}$$

At iteration step k $(k = 1, \ldots, n)$:

Step 1: Put

$$p_k = x_k^{(k)} - x_k^{(k-1)} \quad . \tag{A.2.20}$$

Step 2: Determine by $n-k+1$ line searches implemented in parallel starting from $x_k^{(k)}, \ldots, x_k^{(n)}$ in the common direction p_k the $n-k+1$ step lengths

$$a_k^{(i)} \in \mathbb{R} \qquad (i = k, k+1, \ldots, n) \tag{A.2.21}$$

and in parallel for $i=k, k+1, \ldots, n$ assign

$$x_{k+1}^{(i)} = x_k^{(i)} + a_k^{(i)} p_k \quad . \tag{A.2.22}$$

The equivalence of the search directions of the procedures of Sloboda and Gram–Schmidt follows using (A.2.12) from

$$x_k^{(i)} - x_k^{(k-1)} = \hat{p}_i^{(k-1)} \qquad k = 1, \ldots, n;\ i = k, \ldots, n. \tag{A.2.23}$$

Because of the uniqueness of the minimum points of quadratic functions on lines and the procedure used in (A.2.20) to obtain the search directions p_k, Step 2 can be substituted by

Step 2a: Determine by $n-k+1$ line searches implemented in parallel starting from the points $x_k^{(k-1)}, x_k^{(k+2)}, x_k^{(k+2)}, \ldots, x_k^{(n)}$ and all in the common direction p_k, the $n-k+1$ step lengths

$$a_k^{(k-1)},\ a_k^{(k+1)},\ a_k^{(k+2)}, \ldots, a_k^{(n)} \in \mathbb{R} \tag{A.2.24}$$

and in parallel assign

$$x_{k+1}^{(k)} = x_k^{(k-1)} + a_k^{(k-1)}\, p_k \tag{A.2.25}$$

and for $i = k+1, \ldots, n$,

$$x_{k+1}^{(i)} = x_k^{(i)} + a_k^{(i)}\, p_k \quad . \tag{A.2.26}$$

The connection between the iteration points and search directions is clarified by Fig. 2.

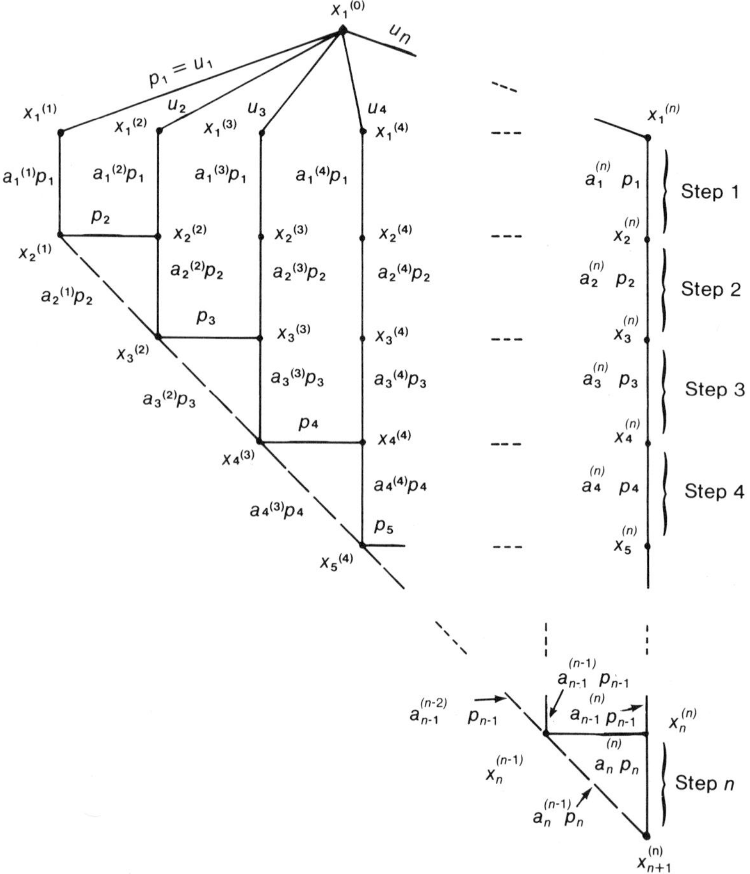

Fig. 2 – Illustrating algorithm 2.

By (A.2.22) and (A.2.25) the following hold for the final iteration point:

$$\begin{aligned} x_{n+1}^{(n)} &= x_1^{(n)} + \sum_{i=1}^{n} a_i^{(n)} p_i \\ &= x_1^{(0)} + \sum_{i=1}^{n} a_i^{(i-1)} p_i \quad , \end{aligned} \tag{A.2.27}$$

that is, $x_{n+1}^{(n)}$ is obtained by a sequence of n line searches in the directions p_1, $\ldots, p_n$ beginning at the initial point $x_1^{(n)}$, or the equivalent with $x_1^{(0)}$. Because of the conjugacy property of the search directions the global minimum point x^* of f is obtained at $x_{n+1}^{(n)}$ with exact line search. From the equality of the search directions it also follows that Sloboda's procedure, starting from the same point as the conjugate Gram–Schmidt procedure, computes the same sequence of minimum points $\{x_1^{(0)}, x_2^{(1)}, \ldots, x_{n+1}^{(n)}\}$ as the sequence $\{x_1, x_2, \ldots, x_{n+1}\}$ computed by conjugate Gram–Schmidt.

For the minimisation of a function f which is not quadratic by the conjugate Gram–Schmidt, or Sloboda, procedures the cycles described in Algorithms 1 and 2 of n iteration steps are repeated until a sufficiently close approximation to a global minimum point of f is achieved. The last iteration point x_{n+1}, or $x_{n+1}^{(n)}$ calculated in a cycle then serves as the starting point for the next cycle. The conjugate Gram–Schmidt and Sloboda procedures are thus restart procedures.

A parallel minimisation procedure which is not a restart procedure is the procedure of Chazan and Miranker. When f is quadratic this constructs the same search directions as the conjugate Gram–Schmidt.

Algorithm 3

An arbitrary starting point

$$x_1^{(0)} \in \mathbb{R}^n \tag{A.2.28}$$

is given together with n linearly independent vectors

$$u_1, u_2, \ldots, u_n \in \mathbb{R}^n . \tag{A.2.29}$$

Start: For $i = 1, \ldots, n$ set

$$x_1^{(i)} = x_1^{(i-1)} + u_i \quad . \tag{A2.30}$$

At iteration step k $(k=1, 2, \ldots)$:

Step 1: Put

$$p_k = x_k^{(1)} - x_k^{(0)} \quad . \tag{A.2.31}$$

Step 2: Determine by n line searches implemented in parallel starting from $x_k^{(1)}, \ldots, x_k^{(n)}$ in the common direction p_k the n step lengths

$$a_k^{(i)} \in \mathbb{R} \qquad (i = 1, \ldots, n) \tag{A2.32}$$

and in parallel for $i = 1, \ldots, n$ assign

$$x_{k+1}^{(i-1)} \doteq x_k^{(i)} + a_k^{(i)} p_k \quad . \tag{A.2.33}$$

Step 3: Put

$$x_{k+1}^{(n)} = x_{k+1}^{(n-1)} + u_j \tag{A.2.34}$$

where

$$j = ((k-1) \bmod n) + 1 \quad . \tag{A.2.35}$$

It can be shown (see [44]) that when f is quadratic exact line search with (A.2.12) makes the following hold:

$$x_k^{(i)} - x_k^{(i-1)} = \hat{p}_{(k-1+i)}^{(k-1)} \quad , \quad k = 1, \ldots, n; \; i = 1, \ldots, n-k+1, \tag{A.2.36}$$

and in the special case of (A.2.13)

$$p_k = x_k^{(1)} - x_k^{(0)} = \hat{p}_k^{(k-1)} = \hat{p}_k, \qquad k = 1, \ldots, n \quad . \tag{A.2.37}$$

Because of the construction used Step 2 can be replaced by:

Step 2a: Determine by n line searches implemented in parallel starting from $x_k^{(0)}, x_k^{(2)}, x_k^{(3)}, \ldots, x_k^{(n)}$ in the common direction p_k the n step lengths

$$a_k^{(0)}, a_k^{(2)}, \ldots, a_k^{(n)} \in \mathbb{R} \tag{A.2.38}$$

and in parallel assign

$$x_{k+1}^{(0)} = x_k^{(0)} + a_k^{(0)} p_k \quad , \tag{A.2.39}$$

and for $i = 2, \ldots, n$

$$x_{k+1}^{(i-1)} = x_k^{(i)} + a_k^{(i)} p_k \quad . \tag{A.2.40}$$

Fig. 3 explains this.

By (A.2.33) and (A.2.39) we have

$$x_{n+1}^{(0)} = x_1^{(n)} + \sum_{i=1}^{n} a_i^{(n-i+1)} p_i \tag{A.2.41}$$

$$= x_1^{(0)} + \sum_{i=1}^{n} a_i^{(0)} p_i \quad .$$

By the conjugacy of the search directions $p_1, \ldots, p_n$ the global minimum point x^* of the quadratic function (A.2.1) is thus reached at $x_{n+1}^{(0)}$. Moreover, starting with a common initial point and using exact line search, Algorithm 3 generates the sequence $\{x_1^{(0)}, x_2^{(0)}, \ldots, x_{n+1}^{(0)}\}$ of minimum points and this is the same as the sequence $\{x_1, x_2, \ldots, x_{n+1}\}$ given by the conjugate Gram–Schmidt procedure, and the sequence $\{x_1^{(0)}, x_2^{(1)}, \ldots, x_{n+1}^{(n)}\}$ given by the Sloboda procedure.

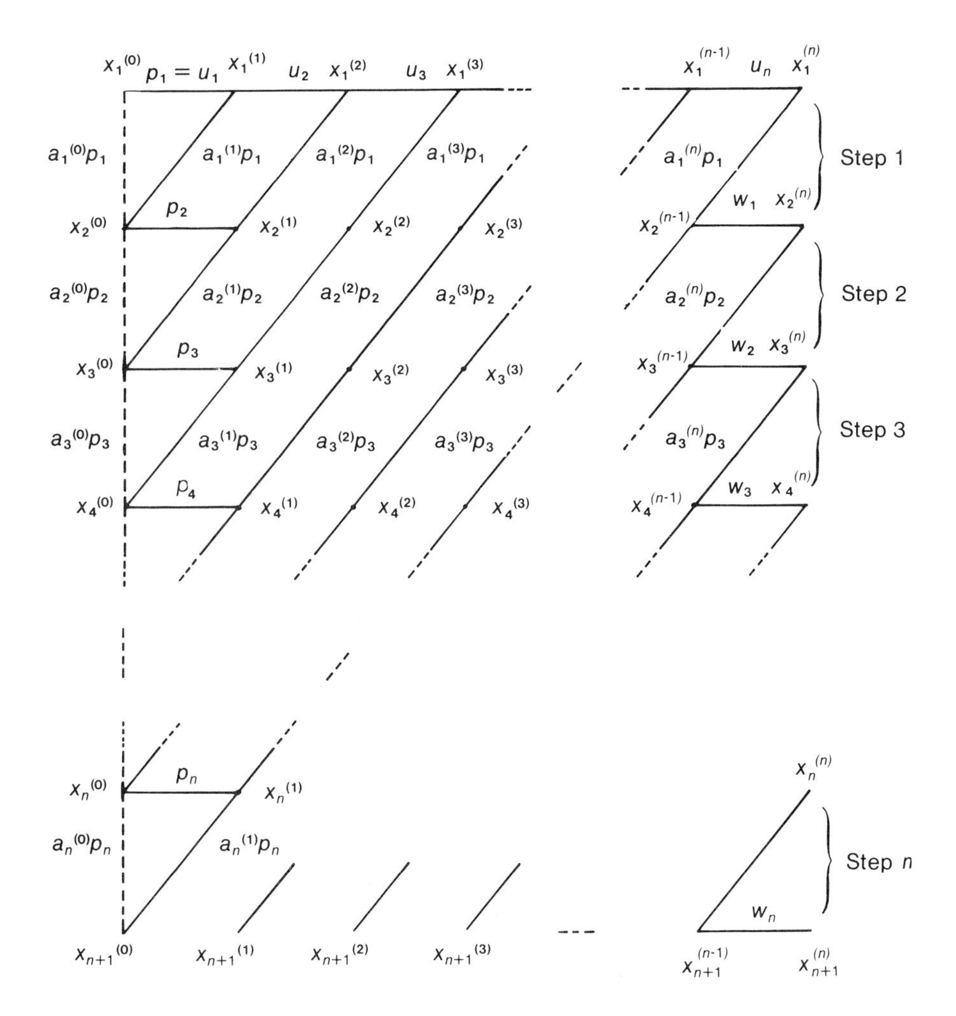

Fig. 3 – Illustrating algorithm 3.

References

[1] Wilkinson, J. H., *Rundungsfehler,* Springer Verlag, Berlin/Heidelberg/New York (1969).

[2] Stone, H. S., Problems of parallel computation, in *Complexity of Sequential and Parallel Numerical Algorithms,* ed. J. F. Traub, Academic Press (1973).

[3] Ibbett, R. N., *The Architecture of High Performance Computers,* Macmillan, London (1982).

[4] Flynn, M. J., Very high speed computing systems, *Proc. IEEE,* **14**, 1901–1909 (1966).

[5] Kupfernagel, K., *Behandlung paralleler Algorithmen für lineare Systeme und deren Auswertung auf einem Parallelrechner.* Diplom-Thesis, Freie Universität Berlin (1979).

[6] Holland, J. H., A univeral computer capable of executing an arbitrary number of subprograms simultaneously, *Proceedings Eastern Joint Computer Conference,* 108–113 (1959).

[7] Kuck, D. J. & Muraoka, Y., Bounds on the parallel evaluation of arithmetic expressions using associativity and commutativity, *Acta Informatica,* **3**, 3, 203–216 (1974).

[8] Heller, D., A survey of parallel algorithms in numerical linear algebra, *SIAM Review,* **20**, 4, October (1978).

[9] Kogge, P. M., Parallel solutions of recurrence problems, *IBM J. Res. Develop.,* **18**, 138–148, March (1974).

[10] Schendel, U. & Brandenburger, J., *Algorithmen zur Lösung Rekurrenter Relationen,* Preprint No. 101/79, Freie Universität Berlin (1979).

[11] Stone, H. S., An efficient parallel algorithm for the solution of a tridiagonal linear system of equations, *J. ACM,* **20**, 1, 27–38, January (1973).

[12] Kuck, D. J., *Structure of Computers and Computations,* John Wiley & Sons, New York (1978).

[13] Kung, H. T., New algorithm and lower bounds, *J. ACM,* **23**, 2(1976).

[14] Abramowitz, M. & Stegun, I., eds., *Handbook of mathematical functions,* Dover, New York (1965).

[15] Chen, S. C. & Kuck, D. J., Time and parallel processor bounds for linear recurrence systems, *IEEE Trans. on Comp.*, **C-24**, 7, 701–717, July (1975).

[16] Borodin, A. & Munro, I., *The Computational Complexity of Algebraic and Numeric Problems,* American Elsevier (1975).

[17] Sameh, A. H. & Brent, R. P., Solving triangular systems on a parallel computer, *SIAM J. Numer. Anal.*, **14**, 6, 1101–1113 (1977).

[18] Householder, A. S., *The Theory of Matrices in Numerical Analysis,* Blaisdell, New York (1974).

[19] Chen, S. C., Kuck, D. J. & Sameh, A. H., Practical parallel band triangular systems solvers, submitted for publication in *ACM TOMS.*

[20] Traub, J. F., Tridiagonal systems on parallel computers, in *Complexity of Sequential and Parallel Numerical Algorithms,* ed. J. F. Traub, Academic Press (1973).

[21] Stoer, J. & Bulirsch, R., *Einführung in die Numerische Mathematik II,* Springer-Verlag Berlin, Heidelberg, New York (1973).

[22] Kuck, D. J. & Sameh, A. H., Parallel computations of eigenvalues of real matrices, *Information Processing,* **71**, 1266–1272 (1971).

[23] Sameh, A. & Kuck, D. J., A parallel QR algorithm for symmetric tridiagonal matrices, *IEEE Trans. Comptrs.*, **C-26**, 147–153 (1977).

[24] Sameh, A. H., On Jacobi and Jacobi-like algorithms for a parallel computer, *Math. Comp.*, **25**, 579–590 (1971).

[25] Ward, R. C., The QR algorithm and Hyman's method on a vector computer, *Math. Comp.*, **30**, 132–142 (1976).

[26] Schendel, U., Gomm, W. & Weistroffer, R., *Prozeduren zur Simulation einiger Vektoroperationen auf einem seriellen Rechner,* Preprint No. 42/1977, Freie Universität Berlin (1977).

[27] Eberlein, P. J. & Boothroyd, J., Solution to the eigenproblem by a norm reducing Jacobi-type method, *Num. Math.*, **11**, 1–12 (1968).

[28] Watkins, D. S., Understanding the QR algorithm, *SIAM Review,* **24**, 427–440 (1982).

[29] Francis, J. G. F., The QR transformation: a unitary analogue to the LR transformation, *Computing J.*, **4**, 265–271, 332–343, (1961/2).

[30] Reinsch, C. H., A stable rational QR algorithm for the computation of the eigenvalues of an Hermitian tridiagonal matrix, *Math. Comp.*, **25**, 591–597 (1971).

[31] Blumenfeld, M., Doll, J., Göbel, D. & Schendel, U., Numerische Algorithmen bei Matrizen, *Automatisierung, Analyse und Synthese Dynamischer Systeme,* Band 65, Berlin (1974).

[32] Schendel, U., Gomm, W. & Weistroffer, R., *Über die Simultan-Iteration zur Bestimmung der p größten Eigenwerte und zugehörigen Eigenvektoren einer (n,n)-Matrix auf einem Parallelrechner,* Preprint No. 30/1977, Freie Universität Berlin (1977).

[33] Shedler, G. S., Parallel numerical methods for the solution of equations, *Comm. ACM,* **10**, 286–291 (1967).
[34] Moore, R. E., *Intervallanalyse,* Oldenbourg Verlag, München, Wien (1969).
[35] Herzberger, J., Some multipoint-iteration methods for bracketing a zero with applications to parallel computation in *Parallel Algorithms for the Efficient Solution of Recurrence Problems,* ed. P. M. Kogge, Digital System. Lab., Stanford Univ., Stanford, CA, Rep. No. 43, September (1972).
[36] de Boor, C., *A Practical Guide to Splines,* Springer Verlag, Berlin, Heidelberg, New York (1978).
[37] Micchelli, C. A. & Miranker, W. L., High order search methods for finding roots, *J. ACM,* **22**, 51–60 (1975).
[38] Gal, S. & Miranker, W. L., Optimal sequential and parallel search for finding a root, *Journal of Combinatorial Theory (A),* **23**, 1–4 (1977).
[39] Winograd, S., On computing the discrete Fourier transform, *Math. Comp.* **32**, 175–199 (1978).
[40] Brigham, E. O., *The Fast Fourier Transform,* Prentice-Hall, New York (1974).
[41] Giloi, W. K., *Rechnerarchitektur,* Springer Verlag, Berlin, Heidelberg, New York (1981).
[42] Müller-Wichards, D. & Gentzsch, W., *Performance comparisons among several parallel and vector computers on a set of fluid flow problems,* Deutsche Versuchs- und Forschungsanstalt für Luft- und Raumfahrt (DVFLR), Report No. IB262–82 RO1, Göttingen, July (1982).
[43] Sloboda, F., Parallel method of conjugate directions in minimization, *Aplikace Matematiky,* **20**, 436–446 (1975).
[44] Chazan, D. & Miranker, W. L., A nongradient and parallel algorithm for unconstrained minimization, *SIAM J. of Control,* **8**, 207–217 (1970).
[45] Hestenes, M., Conjugate direction methods in optimization, *Applications of Mathematics,* No. 12, Springer Verlag, New York–Heidelberg–Berlin (1980).
[46] Elliott, D. F. & Rao, R., *Fast Transforms,* Academic Press, New York (1983).
[47] Schendel, U. & Golm, K., *Vergleich einiger Leistungsdaten von Pipeline-, MIMD- und SIMD-Computern,* Preprint No. 152/1983, Freie Universität Berlin (1983).
[48] Schyska, M., *Parallele Algorithmen in der Nichtlinearen Optimierung,* Diplom-Thesis, Freie Universität Berlin (1983).
[49] Schendel, U. & Schyska, M., *Parallele Algorithmen in der Nichtlinearen Optimierung,* Preprint No. 161/1984, Freie Universität Berlin (1984).
[50] Zakharov, V., Parallelism and array processing, IEEE, *Trans. on Comptrs.,* **C-33**, 45–78 (1984).
[51] Hockney, R. W. and Jesshope, C. R., *Parallel Computers,* Adam Hilger Limited (1981).

Index

Schendel
Introduction to numerical
methods for parallel
computers